CHAPTER ZERO

Fundamental Notions of Abstract Mathematics

CHAPTER ZERO

Fundamental Notions of Abstract Mathematics

SECOND EDITION

Carol Schumacher
Kenyon College

Addison
Wesley

Boston San Francisco New York
London Toronto Sydney Tokyo Singapore Madrid
Mexico City Munich Paris Cape Town Hong Kong Montreal

Sponsoring Editor: *Bill Poole*
Assistant Editor: *RoseAnne Johnson*
Managing Editor: *Karen Guardino*
Marketing Manager: *Michael Boezi*
Manufacturing Manager: *Evelyn Beaton*
Associate Production Supervisor: *Julie LaChance*
Composition: *Windfall Software, using ZzTEX*
Illustration: *LM Graphics*
Cover Designer: *Dardani Gasc Design*
Interior Designer: *Sandra Rigney*
Senior Designer: *Barbara Atkinson*

This text is in the Addison-Wesley Higher Mathematics Series. For more information about Addison-Wesley Mathematics books, access our World Wide Web site at *http://www.aw.com/he/*.

Library of Congress Cataloging-in-Publication Data

Schumacher, Carol.
    Chapter zero : fundamental notions of abstract mathematics / Carol Schumacher.—2nd ed.
        p.   cm.
    Includes bibliographical references and index.
    ISBN 0-201-43724-4
    1. Mathematics. 2. Problem-solving. I. Title.
    QA9 .S376 2000   00-032295
    511.3—dc20

Cover Photo is "Nature Morte avec Bouteille et Cigares" by Juan Gris, 1912. Courtesy of Christie's Images/SuperStock.

Figure 1.3 Copyright © 1977 by Sydney Harris—American Scientist Magazine. Figures 2.1 and 7.2 Copyright © 1996 by Sydney Harris.

Example 6.1.2 from *Induction in Geometry* by L. I. Golovina and I. M. Yaglom, 1979, Moscow: Mir Publishers. Reprinted with permission.

Printed in the United States of America.

1 2 3 4 5 6 7 8 9 10—CRS—03 02 01 00

The first edition of
Chapter Zero
was dedicated to
Robert Eslinger,
who never told me there were *formulas.*

The second edition
is dedicated to my (infinitely patient) daughters,
Sarah and Glynis,
who lost a summer with their Mom
so that it could come to be.

Preface

Students typically begin their undergraduate mathematics education believing that mathematics is a collection of established problem-solving techniques that are "out there" to be learned and applied. But mathematics is more than that: It is a mode of inquiry. If students are to get beyond their superficial view of mathematics, they must change the way they think about the subject. A course that aims to help students turn this corner must be fundamentally different from both earlier and later courses. A different kind of course requires a different kind of book.

Why *Chapter Zero* Was Written

A decade ago at Kenyon College, we in the Mathematics Department were concerned about a perennial problem. Many of our upper-level students were struggling with the basics: proving two sets are equal, showing that a function is one-to-one and onto, handling an induction argument, making sense of equivalence relations and partitions. We decided (like other institutions) that we needed to offer a sophomore-level "introduction to proofs" course that would concentrate on these issues so that they would be less likely to become a roadblock later on. I was given the task of developing and teaching this new course.

I strongly believe that students achieve mathematical maturity by engaging the subject seriously and actively. Students too often are given the impression that they can learn mathematics by watching a professor lecture. Textbooks are written in a way that reinforces this notion. The exercises are relegated to the end of each section, suggesting that the students' work begins only after the reading is done. Furthermore, the students are rarely given responsibility for proving important results that will be used later, suggesting that their work is peripheral to the development of the subject.

I sought a textbook suitable for a course in which the students' own work was the central and indispensable element. Unfortunately, I found that most "transition" textbooks were organized along traditional lines. I therefore decided to write my own. The first edition of *Chapter Zero* was the result, refined by a few years of classroom use at Kenyon College. This second edition corrects some notable flaws (I hope) and draws upon my own and others' experiences using the first edition.

What Sort of Book Is This?

Chapter Zero is a very unusual book. In writing it I have tried to pursue two seemingly contradictory goals. On the one hand, the book contains very few finished proofs. The students who use the book must prove virtually all of the theorems themselves, so that in some ways the book is a long series of interconnected exercises. The students' work is inextricably linked to the development of the subject. On the other hand, I have tried to support the students' developing intuition about mathematics, taking them "backstage" to explore the motivations behind definitions and axioms and giving them practical tips about proof techniques and the construction of arguments. This two-pronged strategy is designed to encourage and enable students to take active responsibility for the mathematics they are learning.

In *Chapter Zero* new ideas are usually presented first in an informal discussion before a formal definition is given. The definition is most often immediately followed by accessible exercises that help the students understand the structure of the definition by constructing or verifying examples or by answering questions about the definition. Other problems and theorems of varying degrees of difficulty follow. The book does not shy away from asking students to prove harder theorems, but it provides guidance ranging from one-line hints to extensive proof outlines that nevertheless leave the details for the students to fill in. Students are even occasionally asked to formulate definitions and theorems themselves.

I have adopted a conversational tone in the book to the extent that mathematical rigor allows. I hope that this style will draw the students into the business of the book and leave them with a good intuitive understanding of the mathematical ideas.

What Is in This Book?

- The essay in Chapter 0 sets the stage for the rest of the book by telling the students something about the mathematical enterprise and describing the philosophy of the book.

- Logic is presented as an instrument for analyzing the content of the mathematical assertions and a tool for constructing valid mathematical proofs and not as an end in itself.

- The treatment of equivalence relations is motivated by a look at partitions. The definition of equivalence relations then emerges from an exploration of the general connection between relations on a set A and collections of subsets of A.

- Sequences are introduced as a special class of function and are used extensively in the chapters on cardinality and the real numbers. In Chapter 7, infinite sets are naturally characterized as those sets containing sequences of distinct terms and countable sets are those whose elements can be listed as a sequence of distinct terms. In Chapter 8, the discussion of the real number system culminates in the study of sequence convergence and completeness.

- After the foundational work in Chapters 0-5, Chapters 6, 7, and 8 are independent of one another and offer three different avenues for continued study. Each develops a general theme and concludes with an important result or idea. Chapter 6, "Elementary Number Theory," focuses on divisibility, first in the integers and then in the integers modulo n, culminating in the result that division is well-defined in $\mathbb{Z}_p$. Chapter 7, "Cardinality," deals with comparing sizes of infinite sets, including Cantor's diagonalization proof and the Schroeder-Bernstein theorem. This is followed by a short discussion of the Continuum Hypothesis and its independence of the standard axioms of set theory. Chapter 8, "The Real Numbers," is an extended exploration of the axioms of the real number system, emphasizing the desirability and necessity of those axioms. In the final section it is shown that every Cauchy sequence of real numbers converges.

- Although the discussion of set theory in the body of the text is informal, Russell's paradox is presented as a rationale for an axiomatic approach. Appendices A and B are a more rigorous treatment. They begin with a presentation of standard axioms for set theory and continue through a set-theoretic construction of the real numbers. The appendices are written in the same style as other chapters in the book, though the development is necessarily less detailed. These may be used as outlines to guide student work on the topics.

New in the Second Edition

Writing a second edition of a book is gratifying, since it gives the author a chance, so to speak, to atone for the sins she committed in the writing of the first edition. Some of them the author noticed herself, and many others were pointed out by some very perceptive reviewers and other colleagues who had used the book.

Major Structural Changes

- The chapter on mathematical induction has been completely rewritten and placed much earlier in the text.

- I have added a (fairly long) section on graph theory to Chapter 4 on relations. This section, like Chapters 6, 7, and 8, is an independent unit that focuses on a small piece of a much larger mathematical subject. It explores the ideas necessary for a discussion of map coloring. Although the section may be omitted, it is an ideal place for students to practice mathematical induction.

Substantial Reworkings

- The chapter on logic has been rewritten to make it a better foundation for mathematical proof. The core philosophy of the chapter, which is to build on students' previous intuition about truth and falsehood in mathematics, is embodied in a new "thought experiment" at the beginning. Students evaluate a list of mathematical statements to determine whether they are true or false, and try to prove their answers.

  Targeted discussions, expanded and improved from the first edition, help to bring the following ideas into focus:

  - learning from truth tables,
  - negating implications,
  - existence theorems,
  - uniqueness theorems,
  - the role of example and counterexample in mathematical proof.

  The chapter as a whole is enhanced by more sample proofs and more opportunities for the students to work on proofs of their own.

- In the chapter on sets, I have expanded and clarified the discussions of set notation, general indexing sets, and power sets. Venn diagrams now make an appearance, as well.

- Though there are several notable improvements to the chapter on functions, the most extensive renovations have occurred in the section on sequences. The discussion of subsequences and subsequence notation is more detailed and considerably clearer. There is, in addition, a fairly thorough introduction to the uses of mathematical induction in the recursive definition of subsequences.

Systematic Improvements

For the second edition I have added new problems and exercises, including many problems of intermediate difficulty and additional exercises in which students work with specific examples. To keep from obscuring the development of the key mathematical ideas, lots of these new problems have been located at the ends of chapters.

> Mathematical asides are included in boxes throughout the text.

Also at the chapter-ends are a few "Questions to Ponder." These may be more difficult problems, open-ended questions, philosophical issues, or foreshadowings of future developments. Some are quite hard. Some are unsolved. Students who wish to delve more deeply are invited to mull over the "Questions to Ponder."

How I Have Used This Book in Classes

At Kenyon College, classes using *Chapter Zero* have been conducted seminar-style, with few lectures by the professor. The professor normally acts as a "moderator" while the students present exercises and proofs to one another. The professor answers questions and tries to steer the discussion into productive channels. Thus, *the most important lines of communication are those between the students.*

Ordinarily, the work is presented by volunteers and critiqued by the other students. The participation of students by presenting and critiquing work in class forms a large part of their grade for the course. The rest of the grade is determined by submitted written work, including take-home exams that ask students to prove theorems that are new to them. Written work is graded both on mathematical content and on clarity and style.

Instructor's Resource Manual

Because this is an unusual book, it requires an unusual classroom approach. For this reason an Instructor's Resource Manual is available. This manual gives teachers additional, more detailed information about the organization of the book and suggestions for its classroom use based on my own experience and that of the colleagues who have used it. A copy of the Instructor's Resource Manual may be obtained from Addison-Wesley or downloaded from *www.aw.com/schumacher*.

Acknowledgments

Many people contributed to the writing of this book. I cannot name them all individually, but I am deeply grateful for their help.

I would especially like to note the contributions of students that have used *Chapter Zero* in a course. Because of their input (direct or indirect) the book is far more usable, easier to read, and more complete. Though all the students who have used the book have contributed to its development, I would especially like to acknowledge the contributions of Stacy Bear, Bill Birchenough, Robin Blume-Kohout, Christine Breiner, Jim Carlone, Cathryne Carpenter, Teena Conklin, Karen Downey, Brent Ferguson, Colleen Hopkins, Jay Ireland, Adrian Polit, Trent Stanley, Brian Vannoni, and Amy Wagaman.

I have found invaluable the professional insights of mathematicians Benjamin Ford, Robin Lerch O'Leary, Dana Mackenzie, Bruce Reed, Edward Saff and his graduate students, Stephen Slack, and Tara Smith, who provided me with much assistance and insight prior to the first appearance of *Chapter Zero*. Kenneth Valente, Jerry Bradford, Patrick Brewer, Marcelo Llarull, and Myrtle Lewin carefully reviewed the first edition; their comments were invaluable to the editing process. Dwayne Collins and Michael Westmoreland, made excellent contributions to both the first and the second editions. Special acknowledgment goes to Deanna Caveny, Dwayne Collins, Myrtle Lewin, and Michael Westmoreland who suggested specific problems that are included in the second edition. Deanna Caveny suggested Exercises 5.1.14 and 5.1.15, as well as Problems 1 and 2 at the end of Chapter 5. Credit for Problem 1.8.7 goes to Dwayne Collins. Myrtle Lewin is responsible for Exercise 1.7.1. Myrtle, who gives out a list of "juicy questions" for her students to think about, was also the inspiration for the inclusion of "Questions to Ponder." Michael Westmoreland recommended Problem 4 at the end of Chapter 2. Though it may sound paradoxical, the additions and changes that these colleagues have suggested have broadened the scope of the book and at the same time helped to sharpen its focus. They have pointed out specific weaknesses in the first edition and have made concrete suggestions for improvement, many of which are incorporated in the second edition. I cannot thank them enough for their help.

Though I suspect he will be surprised to find himself acknowledged for contributions to a mathematics textbook, I must extend a special word of thanks to Professor William McCulloh (Professor Emeritus of Classics, Kenyon College) for his help with Greek etymology.

Naturally, none of this would have been possible without the support of Addison-Wesley. Thanks go to Laurie Rosatone for the confidence she has shown in the project over the years and for generous help she has provided at every step of the way. Special thanks, also, to Carolyn Lee-Davis for her gentle guidance during the preparation of the second edition. Kathy Manley, Ranjani Srinivasan, Susan London-Payne, and David Dwyer were a tremendous help on the first edition. Julie LaChance, RoseAnne Johnson, and Paul Anagnostopoulos have been a marvelous team to work with on the second. You would not be holding this book were it not for their collective efforts. In addition, countless people whose names I don't know have played a part in the production of both editions of *Chapter Zero*: copyeditors, designers, artists, compositors. My thanks to all of them for their tireless efforts and for their meticulous work in trying to make this the best book it can be.

Finally, I extend my very special thanks to my husband, Benjamin Schumacher, who has spent many hours helping me polish and smooth the edges of the book. He has made many suggestions that have improved both the first and second editions of *Chapter Zero*. I would especially like to acknowledge his help with the introductory essay, the chapter on logic, and the appendices on axiomatic set theory which he has virtually co-authored.

Carol S. Schumacher
(SchumacherC@Kenyon.edu)
Gambier, Ohio
Fall, 2000

A Note to the Student:
What This Book Expects from You

What this book expects from you is your active participation. I recently asked Amy, a student who studied out of *Chapter Zero* last year, if she had any advice for students who would be using the second edition of the book. Here was her response to you:

> *Chapter Zero requires active learning, based on a high level of interaction between you and the book. Concepts build on one another and you need to understand the definitions before you can apply them. When you come to a definition, read the accompanying discussion and think about it, try to rephrase it in your own words and also to come up with examples that illustrate it. This will get you thinking about the topic and will help you understand it. If you still don't understand it, ask for help.*

The "high level of interaction" of which Amy speaks requires that you keep pencil and paper handy when you are reading the book. In addition to reflecting on ideas, you will be called upon to work out exercises as you go along. Items marked "Exercise" are designed to contribute to the reading and are usually straightforward. You should work them as soon as you encounter them. Items marked "Problem" or "Theorem" also need your immediate attention, although you do not necessarily need to solve or prove them at once. Instead, you should make sure you understand what is being asked or asserted and how it fits in with the topic being explored. You can go back and work on the full solutions later. "Theorems" are the big results and (except in a few explicitly stated exceptions) are part of an unbroken chain of reasoning built in the book. "Problems" may be less formal and sometimes call upon intuitive mathematical notions outside of the rigorous development in the text in order to illustrate a concept.

In the book you will find practical tips for approaching certain sorts of proofs. Do not make the mistake of glossing over these. They will be extremely useful and important to you. My advice is to mark them in your book for easy reference and make a mental note of them; you may even wish to keep a list of them somewhere in your notebook.

As you proceed, do not imagine that you are leaving "completed" topics behind. Amy has advice to give here, too:

> *Don't be afraid to refer back and forth between chapters. In fact, it is probably better to refer back and forth. Refresh your mind on old concepts every once in a while by skimming back through the early chapters. This will help in the long run.*

I would add that referring to previous sections is not merely for review: The later encounters with a topic will yield additional insight.

At the ends of the chapters you will also find a number of "Questions to Ponder." They are meant to be thought-provoking and fun to work on. Some foreshadow information that comes later in the book. Others are of a philosophical nature or point to deeper mathematical issues. Some are open-ended and do not have cut-and-dried answers. Not all of these questions will interest everyone, but you are encouraged to "play" with those that interest you. Try to think up questions of your own. Mathematics invites exploration.

Do not be discouraged if your progress through the book seems slow. A few pages of text may represent quite a bit of mathematics. Remember, if you are using the book correctly, what is printed here is only a small fraction of the mathematics that is actually being done. The rest is being supplied by you.

Carol S. Schumacher
(SchumacherC@Kenyon.edu)
Gambier, Ohio
Fall, 2000

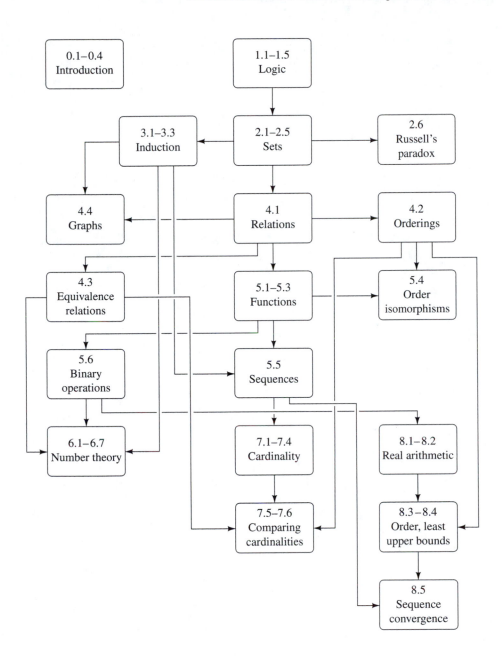

Contents

6   Elementary Number Theory   137

7   Cardinality   157

8   The Real Numbers   179

0 Introduction—An Essay

0.1 Mathematical Reasoning

This is a book about mathematical reasoning. That is, it is a book about the kind of thinking that mathematicians do when they are doing mathematics. Most mathematics courses through the level of elementary calculus teach students how to *use* established mathematical techniques to solve problems. This is a very good beginning, but a complete mathematical education cannot stop there. A student of mathematics must learn to discover and prove mathematical facts on her own. It takes a long time to learn how to create new mathematics.[1] This book is designed for the beginning of the journey.

In mathematical reasoning, logical arguments are used to deduce the consequences (called **theorems**) of basic assumptions (called **axioms**). Most of mathematics is built out of sets; therefore, strictly speaking, the basic assumptions of mathematics are the axioms of set theory (see Appendix A). However, in practice these are rarely invoked directly. More often a mathematician is working from secondary assumptions, often definitions, which could in principle be justified using set theory. Whether the assumptions are really fundamental axioms or merely secondary axioms and definitions, they provide the starting point for careful chains of logical reasoning leading to theorems. All of this is done using a specialized dialect of English,[2] a "mathematical language" designed to describe the ideas of mathematics precisely and unambiguously.

---

[1] By "new" mathematics I do not necessarily mean things never before discovered by anyone. "New" simply means that it is not known to the discoverer—either it was never seen before or it has been forgotten, and so it is reinvented when the need arises. Mathematicians do a lot of this!

[2] Or Russian, or Japanese, or whatever.

0.2   Deciding What to Assume

A brief digression is in order. Familiar to most are Euclid's five axioms of plane geometry, developed about 300 B.C. In modern language they are as follows:

1. Two points determine a line.

2. A line segment can be extended indefinitely.

3. A point and a radius determine a circle.

4. Any two right angles are equal.

5. Given any line and any point not on that line, there is one and only one line through that point that is parallel to the first line. (This version of axiom 5 is actually due to John Playfair (1748–1819). Euclid's original is a bit more complicated.)

From these five statements, Euclid was able to deduce rigorously all of the geometry known in his day, and then some. The axiomatic method was so successful that it became a model of mathematical thinking. Euclid's *Elements* was used as a geometry text for over 20 centuries, and even today most students of geometry learn it using his general approach.

So Euclid achieved a lot. But how did he decide what to start with in the first place? Euclid and geometers for centuries after him believed that his axioms embodied a kind of absolute truth about the plane. He had intuitive notions of planes, lines, and so forth, and he wrote down axioms that he thought described them. Geometers liked Euclid's first four axioms, but there was a lot of controversy about the fifth axiom. (No one doubted that it was true, but many people figured that it was not required as a basic assumption. They thought that it could be proved from the first four axioms.) To make a long story short, in the first half of the nineteenth century, non-Euclidean geometries were invented. That is, perfectly self-consistent systems of geometry were developed based on axioms 1–4 and a new axiom about parallel lines that contradicted Euclid's axiom 5. For example, the new axiom might be:

5′. Given any line and any point not on that line, there is *more than one* line through that point parallel to the first line.

The geometry derived from the new set of axioms is called non-Euclidean geometry; it can be interpreted as the geometry of a curved surface, with certain curves in the surface playing the role of "lines." After all, "plane," "line," and "point" are *undefined* terms in geometry, Euclid's interpretation of them notwithstanding. So—are Euclid's axioms absolutely true, as was supposed?

Yes and no. Actually, this is the wrong question. In the aftermath of the discovery of non-Euclidean geometry, mathematicians came to see that the choice of mathematical axioms is *completely arbitrary* as long as that collection of axioms is self-consistent (that is, it does not give rise to contradictions).

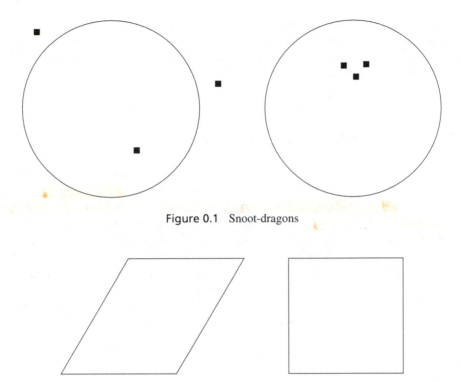

Figure 0.1   Snoot-dragons

Figure 0.2   Square?—Square!

It is perfectly possible to imagine a universe that obeys the laws of a non-Euclidean geometry.[3] Thus different sets of axioms can have equal mathematical validity. On the other hand, this does not really mean that any given consistent set of axioms is as good as any other. When we adopt a particular set of axioms, we usually have some purpose in mind. Euclid chose his axioms to embody his intuitive understanding about plane geometry. The axioms of set theory express our informal notions about sets. In Chapter 8 we will assemble a set of axioms that describe the real numbers. In each of these cases, the axioms capture an abstract idea and make it subject to rigorous mathematical analysis.

The same is true of definitions. They are arbitrary but are chosen with care. For example, in plane geometry we might define a *snoot-dragon* to be "a circle and three points not lying on the circle" (Figure 0.1), but such an animal would probably not be very useful. Then again, we might naively define a *square* to be "a four-sided, equilateral polygon," but we would quickly see that such a figure need not have right angles. If we are actually only interested in the right-angled figures, then we could refine our definition to be "a four-sided, equilateral polygon with four right angles" (Figure 0.2). Furthermore,

---

[3] In fact, the general theory of relativity posits that the geometry of the physical universe is non-Euclidean.

the word "square" already has this widespread conventional meaning. Communication and convenience (and good manners) make it wise to follow established mathematical conventions. Definitions, like axioms, are devised to capture useful notions while conforming to standard practice.

0.3   What Is Needed to Do Mathematics?

Axioms and definitions are the starting point. Once we have settled on the basic assumptions, we can go someplace: We can begin to prove theorems. A **proof** is a (possibly quite long) chain of logical inferences by which we deduce a mathematical statement from the basic assumptions. In practice, we seldom exhibit the entire chain of reasoning all the way from the axioms to the statement we are trying to prove. It suffices to show that this statement follows from previously proved theorems. For once we have proved a theorem, its validity is established just as securely as that of the axioms themselves.

The problem with long chains of reasoning is that one bad link in a chain breaks it (Figure 0.3). A supposed "proof" with an incorrect step in it is not a proof at all (although it may be possible to repair the mistake and make it into a proof). It is therefore crucial to avoid errors at every step.

"I THINK YOU SHOULD BE MORE EXPLICIT HERE IN STEP TWO."

Figure 0.3   One bad link . . .

The primary defense against faulty proofs is strict adherence to the rules of deductive logic. At each step in an argument, these rules are used to establish some new statement on the basis of the previous ones. The step is invalid if it is not sanctioned by the rules, *even if the new statement happens to be true.* An argument consisting entirely of true statements is not valid if the connecting inferences are not justified by logic. Every mathematician must thus know the rules of deductive logic and use them meticulously.

This is harder than it may sound. The statements of mathematics can become very elaborate, and it requires some skill to take them apart and understand how they fit together. For instance, even the apparently trivial process of negation—saying what it means for a given statement to be false—can be tricky if the statement is complex. Try to work out the negation of this statement: "If people recognize you every time you go to the store, then either you live in a small town or else you are a famous celebrity."

Logic, even rigidly adhered to, is inadequate if the language used is imprecise and ambiguous. Because of this, mathematicians have developed specialized conventions of vocabulary and syntax that are much more precise than colloquial language. Hence mathematicians must be able to understand and use mathematical language.

You will notice many things about mathematical language. Most obviously, it involves an extensive vocabulary different from that used in everyday life, including a great many abstract symbols. The symbols are simply a convenience: It is easier to write "x^2" than "the square of x," and "$x \in A$" is more compact than "x is an element of the set A." In each case, the meaning is the same.

There is much more to mathematical language, however, than its baroque vocabulary and arcane hieroglyphics. It is a very formal and stylized dialect. The same connecting phrases appear over and over: "if and only if," "for every," "there exists," and so on. These phrases are carefully chosen to express logical relations with as little ambiguity as possible. Not only do the phrases themselves matter, but their *arrangement* is also crucial. Reflect on the difference between these two statements: "For every poison there is a chemical that is the antidote." "There is a chemical that is the antidote for every poison."

Everyday language is so imprecise that a beginning student of abstract mathematics may feel as though she were being introduced to a foreign language, even though mathematics is written using a variety of English.[4]

With precise language and rigorous logic, a mathematician can construct valid mathematical arguments. However, participation in the wider mathematical enterprise requires a knowledge of the general mathematical "culture." This culture includes ideas, theorems, and techniques common to virtually all branches of mathematics. Open a book of abstract mathematics and you will find sets, relations, functions, proofs by contrapositive and by contradiction, mathematical induction, the close link between equivalence relations and partitions—in short, a core of ideas that every mathematician knows and uses.

---

[4] Or French, or Chinese, or whatever.

0.4   *Chapter Zero*

This book is designed to guide students through an introduction to logic, mathematical language, mathematical proof, and mathematical culture. But it is the *students' journey,* not the author's or the instructor's. It is a truism that the only way to learn mathematics is to do mathematics. So this book is organized like a road map. It lays out the route, and it indicates and explains the significance of landmarks along the way, but it leaves the actual travel to the reader. That is, the reader will be asked to provide proofs for virtually all of the theorems and to construct many of the examples. But there will be help throughout the book in the form of well-motivated definitions, a few sample proofs, discussions of techniques, hints for harder proofs, warnings about possible pitfalls, and generous comments regarding how to think about the ideas.

This book is deliberately incomplete. The reader's active participation is required to fill in the gaps. In fact, the reader must do what a mathematician does when reading any mathematical text: take apart definitions and theorems to see "what makes them tick," fill in the details of sketchy proofs, and put it all back together to grasp the whole. This interactive approach to reading mathematics is an indispensible skill.

Chapter Zero, as its name implies, is essentially a preparation for greater things. After mastering it, a student should be able to pick up virtually any undergraduate textbook of abstract mathematics and read it. A book on algebra, analysis, or graph theory will still be more challenging to read than a mystery novel. It always is. But the student will have the background to understand the ideas and will have practiced the mental discipline of interacting with a mathematical text.

1 Logic

The Introduction says that a theorem is a logical consequence of a collection of axioms; within the context of those axioms it is a true mathematical statement. Our goal in this chapter is to say exactly what we mean by all of these words and to begin to see how to prove theorems.

1.1 True or False?

This section is a preliminary "thought experiment." Its goal is to help you think very explicitly about your own intuition regarding truth and falsehood in mathematics and your intuitive understanding of what is meant by proof. Then we want to build upon that understanding—to tame it, systematize it, and make it into a tool for rigorous mathematical thinking.

I recommend that you compare your own work in this section (your answers and your reasoning) with the work of a fellow student.

Thought Experiment: True or False?

Below you will find a number of mathematical assertions. Most of them deal with arithmetic, algebra, or geometry, since these subjects form a likely base of "common knowledge" for the readers of this book. They are in no particular order. Some statements are true and some are false. Sometimes you will have to assume a context that has not been spelled out for you.

Your goal is figure out the meaning of each statement and then to determine whether it is true or false. Try to justify your answers by a convincing argument, one that would convince a hardened skeptic. (You should be your own harshest critic.) Work with pencil and paper. Keep notes.

I ought to warn you that some of these statements are easy to settle, some are harder, and in one case the answer is unknown. (You are unlikely to resolve this one, but if you do, don't keep the answer to yourself!) Try not to get bogged down in any one problem. There are plenty to keep you busy.

1. The points $(-1, 1)$, $(2, -1)$, and $(3, 0)$ lie on a line.
2. If x is an integer, then $x^2 \geq x$.
3. If x is an integer, then $x^3 \geq x$.
4. For all real numbers x, $x^3 = x$.
5. There exists a real number x such that $x^3 = x$.
6. $\sqrt{2}$ is an irrational number.
7. If $x + y$ is odd and $y + z$ is odd, then $x + z$ is odd.
8. If x is an even integer, then x^2 is an even integer.
9. Every positive integer is the sum of distinct powers of two.
10. Every positive integer is the sum of distinct powers of three.
11. If x is an integer, then x is even or x is odd.
12. If x is an integer, then x cannot be both even and odd.
13. Every even integer greater than 2 can be expressed as the sum of two prime numbers.
14. There are infinitely many prime numbers.
15. For any positive real number x there exists a positive real number y such that $y^2 = x$.
16. Given three distinct points in space, there is one and only one plane passing through them.

Look back over your work. You will probably find that some of your arguments are sound and convincing while others are less so. In some cases you may "know" the answer, but may be unable to justify it—that's OK (for now). Divide your answers into four categories: (Most students will have answers in all four categories.)

   a. I am confident that the justification I gave is conclusive.
   b. I am not confident that the justification I gave is conclusive.
   c. I am confident that the justification I gave is *not* conclusive. (If you gave no justification at all, your answer falls into this category.)
   d. I could not decide whether the statement was true or false.

A number of these problems will be discussed in the coming pages, some will not. But you should keep them in mind as you read. Look back over your notes from time to time. Think about how your work in this section connects with the ideas discussed in the rest of the chapter. Revise your arguments; update the information, as you can. The point is that the logical principles we are about to discuss are not completely alien to you. You already have some intuition about logic, truth and falsehood, and proof. Your goal in this chapter should be to incorporate the systematic treatment of logical principles into your existing understanding, sharpening insights that are right and correcting false impressions, where necessary.

1.2 Statements and Predicates

A statement is a sentence that is either true or false, but not ambiguous. For example:

- George Washington was the first President of the United States.
- Bicycles have six wheels.
- The 10^{47}th digit of π is 7.

These are all statements. Each sentence is either true or false; there is no possible ambiguity. It is not necessary that the truth or falsehood of the statement be known, only that it be unambiguous.

The following are examples of sentences that are not statements.

- *How are you doing?* It makes no sense to ask whether this sentence is true or false; questions have no truth value. Neither do imperative sentences such as "Do your homework." Only declarative sentences have truth value.

- *Picasso's* Les Demoiselles d'Avignon *is an obedient painting.* Sometimes there is no agreed-upon criterion for the truth or falsehood of a sentence. As far as I know, there is no accepted definition of "obedience" that makes sense when applied to a painting.

- *He was six feet tall.* Sometimes the sentence does not provide enough information to be unambiguously true or false. "George Washington was six feet tall," *is* a statement.

1.2.1 EXERCISE

Give some examples of sentences that are statements and some examples of sentences that are not statements.   □

All of these examples bring up some very important issues. Contrast the following sentences:

- *Picasso's* Les Demoiselles d'Avignon *is an obedient painting.*
- *Picasso's* Les Demoiselles d'Avignon *is a beautiful painting.*

Most students agree that neither sentence is a statement. When asked why the first one is not, they usually say that the sentence is absurd, meaningless; therefore, asking whether it is true or false makes no sense. The second sentence is less wacky, but most still reject it as a statement because "beauty is in the eye of the beholder." The truth or falsehood of the sentence is a matter of personal opinion and is thus ambiguous.

Interestingly, a mathematician would say that the two sentences are not statements for *exactly the same reason*. The difficulty in each case lies in the lack of a careful, unambiguous definition for a key phrase— "obedient painting" in one case and "beautiful painting" in the other. As I said in the Introduction, mathematical language must describe the ideas *precisely and unambiguously*. Clear and objective definitions are essential for this. Given the variety of viewpoints on the subject of beauty, it seems impossible to find an *objective* definition of beauty on which everyone could agree—much less a sensible definition of what it means for a painting to be obedient! Without unambiguous definition, we cannot even begin to discuss the unambiguous truth or falsehood of a sentence.

> Our first hint that mathematical English is closely related to, but not really the same as, colloquial English!

Sometimes there are ambiguities in the real world that we deliberately ignore in order to "capture" a subject for mathematical analysis. Consider the sentence "The Earth is not the only planet in the universe inhabited by living creatures." A biologist or a philosopher might say that whether there are living creatures on other planets is *profoundly* ambiguous. The definition of "living creature" is by no means settled and may never be. Does this mean that mathematicians (and biologists and philosophers, for that matter) cannot say anything useful about living creatures? Of course not. We use "working definitions" that are specified explicitly: "For the purposes of this study, a 'living creature' is assumed to have the following characteristics . . . " We make useful simplifying assumptions, a practice that is not only acceptable, but is an essential part of applied mathematics. We state our assumptions explicitly and accept the resulting limitations on the conclusions that we draw.

The sentence "He was six feet tall" is also too ambiguous to be a statement. However, we can interpret it as a statement by assigning a particular meaning to the term "he." We might do this by providing additional contextual information. "George Washington lived in the eighteenth century. He was six feet tall." Or we could think of such assertions as "He was six feet tall" and "x is a positive rational number" as many statements at once, one for each possible interpretation of "he" or x. In this case we might think of "he" and x as **free variables** that may be allowed to take on many possible values. For example, with the values $x = 1$, $x = \pi$, and $x = -7$, the sentence "x is a positive rational number" would become:

- 1 is a positive rational number.
- π is a positive rational number.
- -7 is a positive rational number.

These are all statements. A sentence with a free variable in it that becomes a statement when the free variable takes on a particular value is called a **predicate**. There are also predicates with two, three, or more free variables. For instance, "She went to the movies, and he went fishing," "$x^{2y} = 1$," and "$x + y = 3z$" all have more than one free variable.

1.2.2 EXERCISE

Give examples of mathematical predicates that have two and three free variables. □

Once we can recognize statements and predicates, we need to weave them together to form rigorous logical arguments. To do this, we need to analyze the logical relations between statements.

In symbolic logic, statements and their logical relations are represented by abstract typographical symbols. Statements are represented by single letters P, Q, and so on.

- P := "Beethoven wrote nine symphonies."
- Q := "*Joyful Noise* by Paul Fleishman won the 1989 Newberry Medal."
- A := "A pickle is a flowering plant."

It is convenient to borrow the familiar function notation to represent predicates having one or more free variables: T(x), R(a, b), S(someone, she), and so forth. For instance:

- T(x) := "x has wheels."
- R(a, b) := "$a > 2b$."
- S(someone, she) := "Someone said that she went to Europe in the summer of 1993."

We can make statements out of predicates by assigning values to the free variables. If T(x) is a predicate in the free variable x, and we assign the value a to x, then T(a) is a statement.

1.2.3 EXAMPLE

Suppose T(x) := "x has wheels." Then

- T(Airforce One) := "Airforce One has wheels."
- T(grass) := "Grass has wheels." ■

1.3 Quantification

We can turn a predicate into a statement by substituting particular values for its free variables. There are at least two other ways in which predicates with free variables can be used to build statements. We do this by making a claim about which values of the free variable turn the predicate into a true statement. Consider the predicate about real numbers $x^2 - 1 = 0$. Then we can write the following sentences:

- For all real numbers x, $x^2 - 1 = 0$.
- There exists a real number x such that $x^2 - 1 = 0$.

These are both statements, even though each contains the variable x without any specific meaning attached to it. The first statement is false, since $4^2 - 1 \neq 0$, and the second is true, since $(-1)^2 - 1 = 0$. The phrases **for all** and **there exists** are called **quantifiers**, and the process of using quantifiers to make statements out of predicates is called **quantification**.

In mathematical English, when we say "there exists a bludger," we don't imply that there exists *only one* bludger. We mean that there is *at least one*. If we want to say that there is *one and only one* bludger, we have to say "there exists a unique bludger."

1.3.1   EXERCISE

Suppose we understand the free variable z to refer to fish.

1. Give an example of a predicate A(z) for which "For all z, A(z)" is a true statement.

2. Give an example of a predicate B(z) for which "For all z, B(z)" is false but "There exists z such that B(z)" is true.   □

"For all" is called the **universal quantifier** and "there exists . . . such that" is called the **existential quantifier.** They are so common in mathematical language that there are universally recognized symbols to represent them. The symbol for "for all" is ∀, the symbol for "there exists" is ∃, and the symbol for "such that" is ∍. For instance, we might say "∀ positive real numbers y, y has a positive square root" or "∃ a positive integer n ∍ n is even."

These symbols are not generally used in formal writing, so I will not use them again in the text. However, they are very convenient and are used all the time in informal mathematical discourse. My advice is to adopt them for your own use.

Quantification is so important in mathematical language that further remarks are in order. First, notice that we must take care to specify the "universe" of acceptable values for the free variables. If we are talking about real numbers, then "For all x, $x + 1 > x$" is a true statement; but if x could be Elsie the Cow, then matters are not so clear! If there is any possibility of confusion, we will have to state the range of acceptable values explicitly by saying something like "For all real numbers x, $x + 1 > x$." This can be vitally important. Compare, for instance, the statements

- For all positive real numbers x, $x > x/2$, and

- For all real numbers x, $x > x/2$.

It is essential to provide enough context to avoid ambiguities.

Second, you may find it curious that a sentence might contain a variable, as quantified statements do, and yet be a statement. The variables in statements with quantifiers are called **bound variables.** Suppose that x is a free variable taking values in the integers.

- The predicate "$x > 0$" makes an assertion about a single (but unspecified) integer x.
- The statement "For all x, $x > 0$" makes an assertion about *all* integers x, namely that they are all positive.
- Likewise, "There exists x such that $x > 0$" makes an assertion about all integers x, namely that among them there is a positive one.

The first is not a statement, the last two are. When the variables are bound, there is no ambiguity. When the variables are free, the sentence is ambiguous.

If a predicate has more than one free variable, then we can build statements by using quantifiers for each variable. The sentence "$y^2 = x$" is a predicate with two free variables, which we will suppose refer to positive real numbers. From this we could make the statement

For all x there exists y such that $y^2 = x$.

Note that the *order* of the quantifiers matters greatly. The statement

There exists y such that for all x, $y^2 = x$

is quite different—in fact, it is false, while the previous statement is true. (Take a moment to reflect on this; make sure you understand the difference.)

1.3.2   EXERCISE

Consider the following two statements.

1. There exists x and there exists y such that $y^2 = x$.
2. There exists y and there exists x such that $y^2 = x$.

Did quantifying over y first and then x (rather than the other way around) change the meaning of the statement? What if the quantifiers had both been "for all" instead of "there exists"?   □

1.3.3   EXERCISE

Consider the predicate about integers "$x = 2y$," which contains two free variables. There are six distinct ways to use quantification to turn this predicate into a statement. (Why six?) Find all six statements and determine the truth or falsehood of each.   □

It is worth noting that the phrases "for all," "for any," and "for every" are used interchangeably. Though they may convey slightly different shades of meaning in colloquial English, they all mean the same thing in mathematical English. Similarly, we might say "For some positive real number x, $x^3 - 100 > 0$" instead of "There exists some positive real number x such that $x^3 - 100 > 0$."

1.4   Mathematical Statements

We are interested in mathematics, so we will focus on mathematical statements from now on. Since we are not yet working with a specific mathematical context, I will (for the moment) discuss statements whose context and meaning you should be able to provide from your previous mathematical education. (If you occasionally run across one that you don't understand, don't worry. We are not really interested in content here, just form. Read for the general message.)

The vast majority of mathematical statements can be written in the form "If A, then B," where A and B are predicates.

> But wait! If A and B are predicates involving free variables, then surely "If A, then B" is also a predicate. How do I get away with calling it a statement? In fact, by itself it is not. But it is standard practice to interpret the predicate "If A, then B" as a statement, by assuming universal quantification over the variable(s); that is, "If A(x), then B(x)" is interpreted as "For all x, if A(x), then B(x)." We will follow this convention. (I suspect that you, unknowingly, follow the convention yourself—Or did you look at virtually every statement in the "thought experiment" and argue that it was ambiguous because you didn't know the values of the free variables?) We will say more about this a little later.

1.4.1   DEFINITION

A statement in the form "If A, then B," where A and B are statements or predicates, is called an **implication.**

A is called the **hypothesis** of the statement "If A, then B." B is called the **conclusion.**

Here are some examples of implications.

1.4.2   EXAMPLE

1. If $x + y$ is odd and $y + z$ is odd, then $x + z$ is odd.
2. If x is an integer, then x is either even or odd, but not both.
3. If $x^2 < 17$, then x is a positive real number.
4. If x is an integer, then $x^2 \geq x$.
5. If f is a polynomial of odd degree, then f has at least one real root.　■

1.4.3   EXERCISE

Identify the hypotheses and conclusions in each of the implications given in Example 1.4.2.　□

Often mathematical statements that don't appear to be implications really are, since they can be rephrased as implications.

1.4.4 EXAMPLE

1. "$\sqrt{2}$ is an irrational number" is the same as "If $x > 0$ and $x^2 = 2$, then x is irrational."

2. "For all real numbers x, $x^3 = x$" is often written as "If x is a real number, then $x^3 = x$." ∎

Most mathematical statements that are not implications are statements that assert the existence of something—in effect, predicates with existential quantification over the variables. Here are a couple of examples of existence statements.

1.4.5 EXAMPLE

1. There exists a real number x, such that $x^3 = x$.

2. There exists a line in the plane that passes through the points $(-1, 1)$, $(2, -1)$, and $(3, 0)$. ∎

1.5 Mathematical Implication

Since most mathematical statements are implications (that is, they can be written in the form "if A, then B," where A and B are predicates in one or more variables) we will spend considerable time talking about them. I will begin by appealing to your intuition to motivate the definition of what it means to say that an implication is true. Then we will discuss the logical principles governing the truth and falsehood of more complicated statements. Finally we will talk about various methods of proof.

1.5.1 EXAMPLE

If x is an integer, then $x^2 \geq x$.

Proof. If $x = 0$, then $x^2 = x$, so certainly $x^2 \geq x$. The same is true if $x = 1$. If $x > 1$, then $x^2 > 1 \cdot x = x$. If $x < 0$, then $x^2 > 0 > x$. This accounts for all integer values of x. ∎

What exactly did we do when we proved the theorem? We studied all values of the variable x for which the hypothesis "x is an integer" is true and showed that for those cases the conclusion "$x^2 \geq x$" is true also. We didn't consider values of x that were not integers (that is, values of x for which the hypothesis was false). We understood intuitively that those values were irrelevant to our case.

Notice that we have intuitively assumed universal quantification over the free variable.[1] To clarify, we actually proved "For all x, if x is an integer, then $x^2 \geq x$."

More generally, let us assume that $A(x)$ and $B(x)$ are predicates involving the free variable x. Let $P(x)$ be the predicate "If $A(x)$, then $B(x)$." In the discussion above, we considered all values of x for which $A(x)$ is true, and we said that $P(x)$ should be considered to be true if $B(x)$ was true for all those values. We also said that we were uninterested in the truth value of $B(x)$ at values of the variable x that made $A(x)$ false, for those values of x were not relevant to the truth (or falsehood) of $P(x)$. In fact, we dealt, in one way or another, with *all* possible values of x. We have said what it means for the *statement* "For all x, $P(x)$" to be true! "For all x, $P(x)$" is true unless there is at least one value of x for which $A(x)$ is true and $B(x)$ is false. That is, for specific values of x, $P(x)$ is true unless $A(x)$ is true and $B(x)$ is false. In summary, for a given value of x,

$$P(x) \text{ is } \begin{cases} \text{true} & \text{if } A(x) \text{ and } B(x) \text{ are both true.} \\ \text{false} & \text{if } A(x) \text{ is true and } B(x) \text{ is false.} \\ \text{true} & \text{if } A(x) \text{ is false (regardless of the truth value of } B(x) \text{).} \end{cases}$$

1.5.2 EXERCISE

Consider now the slightly different statement "If x is an integer, then $x^3 \geq x$."

1. Show that "If x is an integer, then $x^3 \geq x$" is false.

2. Thinking in terms of hypotheses and conclusions, explain what you did to show that the statement is false.   □

A value of x that makes the hypothesis A true and the conclusion B false is called a **counterexample**. In order to show that an implication is false, all we need to do is to provide *one* such example. We now see what differentiates true implications from false ones. An implication "If $A(x)$, then $B(x)$" is true if $B(x)$ is true whenever $A(x)$ is. The implication is false if there is even one value of the variable for which the hypothesis is true and the conclusion is false.

1.5.3 EXERCISE

Occasionally you will see "If A, then B" written as "A is sufficient for B" or "B is necessary for A" or "B, if A" or "A only if B." Explain why it is sensible to say that each of these means the same thing.   □

We can summarize our discussion of the truth and falsehood of implications with the following table.

---

[1] For simplicity, the predicates that I refer to have only one variable. Parallel statements apply to predicates with more than one variable. Quantification is assumed over all relevant variables.

A	B	*If* A, *then* B
T	T	T
T	F	F
F	T	T
F	F	T

The various lines in the table give all possible combinations of truth values for generic statements A and B. (Of course, for any specific pair of statements the truth values are determined and will therefore lie in a single line of the table.) The final column then gives the truth value for the resulting implication.

For *predicates* A(x) and B(x), "If A(x), then B(x)" is true if for all possible values of x the truth values of A and B fall only in the first, third, or fourth lines of the table. It is false if even a single value of x lands A(x) and B(x) in the second line.

An implication in which the hypothesis is false is often said to be **vacuously true**.

The term "vacuous" is used in the sense of "devoid of meaning." Sometimes statements that are vacuously true seem to us meaningless or even false. Consider, for instance, the statement

> *If the moon is made of green cheese, then chocolate prevents cavities.*

One might think that the statements "The moon is made of green cheese" and "Chocolate prevents cavities" are surely unrelated. Clearly one does not "imply" the other in everyday usage. But since the moon is not made of green cheese, the hypothesis is false, and our formal rules say that the implication is true. The moral is: Though the commonplace ideas about implication are closely related to the mathematical ones, it is important to remember that to a mathematician, implication is a specialized logical relation that need not have anything to do with cause and effect, as it does in everyday usage.

Why then did we define the truth of implications in such a peculiar way? Remember the proof of Example 1.5.1: We considered only integers x because we understood intuitively that the cases in which the hypothesis was false were irrelevant to our situation. It would have been strange to consider the case in which x was a leopard. Such a case *could never generate a counterexample,* so the truth of the implication was not in danger from leopards.

1.6   Compound Statements and Truth Tables

Suppose that A and B are statements. More complex statements can be built from these, and we can examine their logical structure from the point of view of truth or falsehood. We have already studied "If A, then B" in detail. Symbolically, we write "If A, then B" as $A \Longrightarrow B$, which is read **"A implies B."** Recall the table that we used to summarize our discussion of implications:

A	B	$A \Longrightarrow B$
T	T	T
T	F	F
F	T	T
F	F	T

This is an example of a **truth table.** As you remember, each line of the truth table gives all possible truth values for A and B and the resulting truth value of the implication. If A and B are predicates, "If A, then B" is true if all possible values of the free variable(s) make the truth values of A and B fall in the first, third, or fourth line of the table.

In addition to implication, statements A and B can be combined in a number of ways. The most important are in the following list.

- "A and B" is called the **conjunction** of A and B. We denote it by $A \wedge B$.
- "A or B" is called the **disjunction** of A and B. We denote it by $A \vee B$.
- "Not A" is called the **negation** of A. We denote it by $\sim A$.
- "A if and only if B" is called the **equivalence** of A and B. We denote it by $A \Longleftrightarrow B$. ("If and only if" is often abbreviated iff.)

All of these are called **compound statements.** As in the case of implication, the truth values of these compound statements are defined in terms of the truth values of their individual components.

1.6.1   DEFINITION

Suppose that A and B are statements. The following truth table gives the truth values of $A \Longrightarrow B$, $A \wedge B$, $A \vee B$, $\sim A$, and $A \Longleftrightarrow B$ in terms of the truth values of A and B.

A	B	A implies B $A \Longrightarrow B$	A and B $A \wedge B$	A or B $A \vee B$	not A $\sim A$	A iff B $A \Longleftrightarrow B$
T	T	T	T	T	F	T
T	F	F	F	T	F	F
F	T	T	F	T	T	F
F	F	T	F	F	T	T

1.6.2   EXERCISE

Examine the preceding table. Given the colloquial meaning of the terms "and," "or," "not," and "equivalent," explain why the truth values given in the table make sense. (Note that the mathematical "or" corresponds to the colloquial "and/or." If you wish to indicate that one or the other of two statements is true *but not both,* you must say so explicitly.)   □

More complex compound statements can be formed by combining conjunction, disjunction, negation, implication, and equivalence in various ways. Given the basic truth tables presented, we can find the truth tables for other compound statements.

1.6.3   EXAMPLE

Given that A and B are statements, here are the truth tables for

1. $B \wedge \sim B$:

B	$\sim B$	$B \wedge \sim B$
T	F	F
F	T	F

Notice that the statement $B \wedge \sim B$ is *always* false regardless of the truth value for B. There is no need to check values for free variables. No choice of a free variable will ever yield a true statement of the form $B \wedge \sim B$. *(Think about what is meant by $B \wedge \sim B$. Explain why it makes sense for this statement to be false always.)* A compound statement that is always false regardless of the truth values of the simpler statements involved is called a **contradiction**.

2. $(A \wedge \sim B) \Longleftrightarrow \sim(A \Longrightarrow B)$:

A	B	$\sim B$	$A \wedge \sim B$	$A \Longrightarrow B$	$\sim(A \Longrightarrow B)$	$(A \wedge \sim B) \Longleftrightarrow \sim(A \Longrightarrow B)$
T	T	F	F	T	F	T
T	F	T	T	F	T	T
F	T	F	F	T	F	T
F	F	T	F	T	F	T

Just as $B \wedge \sim B$ was false regardless of the truth value for B, notice that $(A \wedge \sim B) \Longleftrightarrow \sim(A \Longrightarrow B)$ is true regardless of the truth values of A and B. A compound statement that is always true is called a **tautology**.   ∎

You can also form compound statements involving three or more simpler statements. Work out the following example of a tautology for yourself.

1.6.4   EXERCISE

Verify that

$$(A \Longrightarrow (B \vee C)) \Longleftrightarrow ((A \wedge \sim B) \Longrightarrow C)$$

is a tautology by showing that

$$(A \Longrightarrow (B \vee C)) \quad \text{and} \quad (A \wedge \sim B) \Longrightarrow C$$

have the same truth values.

A	B	C	B ∨ C	∼B	A ∧ ∼B	A ⟹ (B ∨ C)	(A ∧ ∼B) ⟹ C
T	T	T					
T	T	F					
T	F	T					
T	F	F					
F	T	T					
F	T	F					
F	F	T					
F	F	F					

□

Notice that truth tables for statements involving only one primitive state-ment have only two rows. (See part 1 of Example 1.6.3.) If there are two primitive statements (e.g., part 2 of Example 1.6.3) we use four rows. Ex-ercise 1.6.4 involved three primitive statements and we used eight rows. These were the number of rows necessary to list all possible combina-tions of true and false for the primitive statements. How many rows will you need to work out a truth table for a compound statement involving four or more statements?

We could list combinations of T's and F's in any order, but then we would have to keep track of which have been listed and which not, make sure there were no repetitions, and so forth. It is convenient to have a pattern scheme that is guaranteed to give us what we want without a lot of trial and error. Look back at the examples and notice the patterns of T's and F's used to give all possible combinations of true and false for one, two (Example 1.6.3), and three statements (Exercise 1.6.4). Can you guess what pattern of T's and F's will work if there are four statements? What if there are more than four?

1.7   Learning from Truth Tables

Ultimately, truth tables are not really much good unless they teach us something. This section will, in a few sample lessons, give you a sense of what sorts of things we can learn from truth tables.

For the entire section, the capital letters A, B, and C will be understood to be predicates.

Lesson 1—Tautologies

Since a tautology is true regardless of the truth values of the underlying primitive statements, tautological statements express logical relationships that hold in any context. The following exercise contains some early lessons.

1.7.1   EXERCISE

Consider the following statements.

1. $(A \Longrightarrow (B \wedge C)) \Longrightarrow (A \Longrightarrow B)$.
2. $(A \wedge (A \Longrightarrow B)) \Longrightarrow B$.
3. $((A \Longrightarrow B) \wedge (B \Longrightarrow C)) \Longrightarrow (A \Longrightarrow C)$.

Each of these statements is a tautology and each embodies an important (and fairly intuitive) logical principle.

- Your first task is to verify that the statements are tautological by constructing truth tables for them.

- Your second task is to figure out what the logical principles are and what they tell us about proving theorems. (It will help to convert the symbols into words.)   □

Lesson 2—What About the Converse?

1.7.2   DEFINITION

The implication $B \Longrightarrow A$ is called the **converse** of $A \Longrightarrow B$.

1.7.3   EXERCISE

Construct a truth table to show that it is possible for $A \Longrightarrow B$ to be true while its converse $B \Longrightarrow A$ is false, and vice versa.   □

So what is the moral of this exercise? The truth of the statement "If A, then B" *does not* imply the truth of its converse. That is, knowing that A implies B *does not* tell us that B implies A.

Figure 1.1   Confusing the statement with its converse

1.7.4   EXAMPLE

Construct the converses of the following statements.

    1. If Elsie is a cow, then Elsie is a mammal.

    2. If $x = 0$, then $x^2 = 0$.   ■

Note that "If Elsie is a cow, then Elsie is a mammal" is a true statement, whereas its converse is false. After all, Elsie might be a kangaroo, a mammal that is not a cow. Note, however, that the truth table you constructed in Exercise 1.7.3 does not go so far as to tell us that if $A \Longrightarrow B$ is true, then its converse is false. Sometimes an implication and its converse are both true, as illustrated by the example "If $x = 0$, then $x^2 = 0$."

Moral. You have to treat a statement and its converse as distinct mathematical claims, each of which requires separate verification.

1.7.5   EXERCISE

Find an example of a true statement whose converse is false and one whose converse is true.   □

Lesson 3—Equivalence and Rephrasing

Consider the truth table for $A \Longleftrightarrow B$:

A	B	A $\Longleftrightarrow$ B
T	T	T
T	F	F
F	T	F
F	F	T

Notice that A $\Longleftrightarrow$ B is true exactly when A and B have the *same truth value*. Suppose A and B are predicates and A $\Longleftrightarrow$ B is true; then we say that A and B are equivalent. This is because A and B are either both true or both false. Thus if we manage to prove A, we know that B is true, too. Conversely, if we prove B, we know that A is true. For all mathematical purposes we may view them as *the same statement phrased in different ways.*

Since equivalent statements are just different ways of stating the same idea, we can use truth tables to explore different ways of phrasing certain sorts of mathematical statements.

1.7.6 EXAMPLE

In Exercise 1.6.4 you showed that

$$(A \Longrightarrow (B \vee C)) \quad \text{and} \quad ((A \wedge {\sim}B) \Longrightarrow C)$$

are equivalent statements. In other words, any statement in the form "If A, then B or C" can be rephrased in the form "If A and not B, then C." (Of course, since "B or C" has the same meaning as "C or B," it is easy to see that "If A and not C, then B" is also equivalent.)

This is an important principle because when we want to prove a statement of the form "If A, then B or C," we usually prove one of the two equivalent forms:

- If A and not B, then C.
- If A and not C, then B. ■

1.7.7 EXERCISE

Show by constructing a truth table that

$$(A \Longleftrightarrow B) \Longleftrightarrow ((A \Longrightarrow B) \wedge (B \Longrightarrow A))$$

is a tautology. □

This truth table shows that the statement "A $\Longleftrightarrow$ B" is equivalent to the conjunction of "If A, then B" and "If B, then A." That is, "A $\Longleftrightarrow$ B is true" can be rephrased by saying that "If A, then B" and its converse are both true statements.

This gives us a method for proving that two statements A and B are equivalent. We have to prove two implications. We first prove that A implies B and then that B implies A.

The phrases "A is equivalent to B" and "A if and only if B" are used interchangeably. You will also occasionally see "A is necessary and sufficient for B."

1.7.8 EXERCISE

There are some very useful rephrasings that involve negation. Construct a truth table that will allow you to compare the truth values of the following four statements.

$$\sim(A \wedge B) \qquad \sim A \wedge \sim B \qquad \sim(A \vee B) \qquad \sim A \vee \sim B$$

Which pairs are equivalent? □

We will discuss other rephrasings that involve negation later in the chapter.

1.8 Negating Statements

Consider the truth table for the negation of a statement A.

A	$\sim$A
T	F
F	T

Notice that if A is a predicate, $\sim$A is a predicate that is true exactly when A is false and false when A is true. Thus if we manage to prove A, we know that $\sim$A is false. Conversely, if we *disprove* A, we know that $\sim$A is true. Thus, the negation of A is a statement of what it *means* for A to be false.

We can always write "It is not true that A" for the negation of A, but generally speaking it is more useful, when we are proving theorems, to say what *is* true rather than to say what is *not* true; negative statements do not generally tell us as much as positive statements. So it is important to be able to translate a negative statement into a positive statement. For instance, if x is a free variable taking its values in the integers it is usually preferable to restate "x is not even" as "x is odd."

1.8.1 EXERCISE

Rephrase the statement "x is not greater than 7" in positive terms. □

Though it is often possible to rephrase a negative statement as a positive statement, this is not always the case. (In which case, of course, we have to leave it in negative terms.)

Not only do we need to be able to rephrase simple statements like "x is not even" in positive terms, we need to be able to negate more complicated statements, as well. Since mathematical statements are often implications, since they have quantifiers, conjunctions, and disjunctions, we need to know how to interpret the negations of these in positive terms. The good news is that there are some general rules to help us along.

Exercise 1.7.8 told us how to negate the conjunction and disjunction of two predicates. The negation of A ∨ B is ~A ∧ ~B and the negation of A ∧ B is ~A ∨ ~B.

1.8.2 EXERCISE

Think colloquially about the meaning of AND and AND/OR. Explain why it makes sense for the negation of A ∨ B to be ~A ∧ ~B and for the negation of A ∧ B to be ~A ∨ ~B.

□

1.8.3 EXERCISE

Negate the following statements. Write the negation as a positive statement, to whatever extent is possible.

1. $x + y$ is even and $y + z$ is even. (x, y, and z are fixed integers.)
2. $x > 0$ and x is rational. (x is a fixed real number.)
3. Either l is parallel to m, or l and m are the same line. (l and m are fixed lines in $\mathbb{R}^2$.)
4. The roots of this polynomial are either all real or all complex. (A complex root is, for the purposes of this exercise, one that has a nonzero imaginary part. In ordinary mathematical usage, real numbers are also complex numbers, they are just numbers whose imaginary part is zero.)

□

Since quantifiers show up in mathematical statements, we must know how to negate statements containing them. As usual, to negate a statement we must decide what it means for the statement to be false. Consider the statement "All senators take bribes." Under what circumstances would this be false? In order to show it to be false, we would have to show that there is (at least) one senator who does not take bribes. The negation of the statement "All senators take bribes" is the statement "There exists some senator who does not take bribes."

1.8.4 EXERCISE

Using similar reasoning, find the negation of the statement "There exists a fast snail." □

1.8.5 EXERCISE

Negate the following statements. Write the negation as a positive statement, to whatever extent is possible.

1. There exists a line in the plane passing through the points $(-1, 1)$, $(2, -1)$, and $(3, 0)$.
2. There exists an odd prime number.
3. For all real numbers x, $x^3 = x$.
4. Every positive integer is the sum of distinct powers of three.
5. For all positive real numbers x there exists a real number y such that $y^2 = x$.
6. There exists a positive real number y such that for all real numbers x, $y^2 = x$. □

1.8.6 EXAMPLE

Sometimes it helps to deal with very complex statements more carefully, one step at a time. Consider the statement "All Martians are short and bald, or my name isn't Darth Vader."

We now consider various substatements:

- A := All Martians are short and bald.
- B := My name isn't Darth Vader.
- C := All Martians are short.
- D := All Martians are bald.

Clearly, A is equivalent to C $\wedge$ D. Our original statement is A $\vee$ B. So the negation of our original statement is

$$\sim(A \vee B) \Longleftrightarrow (\sim A \wedge \sim B) \Longleftrightarrow (\sim(C \wedge D) \wedge \sim B) \Longleftrightarrow (\sim C \vee \sim D) \wedge \sim B.$$

We can now clearly see that the negation of "All Martians are short and bald or my name isn't Darth Vader" is "Either some Martian is tall or some Martian has hair, and my name is Darth Vader." ∎

1.8.7 PROBLEM

If you want a challenge, try using this process to negate

You can fool some of the people all of the time, and some of the people none of the time, but you cannot fool all of the people all of the time.

□

1.8.8 EXERCISE

Let A and B be statements. Show by constructing a truth table that the following statements are equivalent:

$$\sim(A \Longrightarrow B) \quad \text{and} \quad A \wedge \sim B.$$

□

1.8.9   EXERCISE

Negate the statement "If $x^2 > 14$, then $x < 10$." Using your intuition about this example to help you, explain why it makes sense to say that if A is true and B is false, then "If A, then B" is false.   □

Negating a statement of the form "If $A(x)$, then $B(x)$" has a slight wrinkle. If $A(x)$ and $B(x)$ are predicates, $A(x) \wedge \sim B(x)$ is not a statement at all, it is a predicate, too. It seems bad to negate a statement and get a predicate! The solution to this conundrum lies in the fact that "If A, then B" is not really a statement, either. The *statement* is actually "For all x, if $A(x)$, then $B(x)$." So when we negate the implication, we have to negate the quantifier, as well. The negation of the statement "For all x, $A(x) \Longrightarrow B(x)$" is, therefore, "There exists x such that $A(x) \wedge \sim B(x)$." But following the convention that suppresses the universal quantifier in the implication, the existential quantifier in its negation is often suppressed, as well—but only if there is no possibility of confusion as a result! Absolute clarity is always the goal; if there is any ambiguity, always include the quantifier.

This is a good time to review the discussion of counterexamples on page 16. There we provided an intuitive discussion of this idea. Notice that our intuitive discussion of what it means for an implication to be false imposed an existential quantifier.

1.8.10   EXERCISE

Negate the following statements. Write the negation as a positive statement, to whatever extent is possible.

1. If x is an odd integer, then x^2 is an even integer.
2. If $x + y$ is odd and $y + z$ is odd, then $x + z$ is odd.
3. If f is a continuous function, then f is a differentiable function.
4. If f is a polynomial, then f has at least one real root.   □

1.8.11   EXERCISE

Let f be a function that takes real numbers as inputs and produces real numbers as outputs. Negate the following statement.

> For all positive real numbers r, there exists a positive real number s such that if the distance from y to 3 is less than s, then the distance from $f(y)$ to 7 is less than r.

(*Hint:* This problem is really tricky; it will help to think carefully about quantifiers.

- The statement is of the form
  > For all r there exists s such that $(A(s, y) \Longrightarrow B(r, y))$.

  This is, in turn, of the form
  > For all r there exists s such that $P(r, s, y)$.

- Since the predicate P(r, s, y) is quantified over both r and s, y is the only free variable in the implication (A(s, y) $\Longrightarrow$ B(r, y)) which must, therefore, be interpreted as "For all y, (A(s, y) $\Longrightarrow$ B(r, y))."
- So to negate the statement properly, we must interpret it as

For all r there exists s such that for all y, (A(s, y) $\Longrightarrow$ B(r, y)).

Does the statement that you come up with for the negation coincide with your intuition about what it would mean for the statement to be false?)　　　□

1.9　Existence Theorems

We are finally ready to discuss proof! We will start with proofs of existence. A theorem that asserts the existence of something is called an **existence theorem.** Recall the following statement from the "thought experiment." I presume that you decided the statement

There exists a real number x such that $x^3 = x$

was true. How is this demonstrated? Well, the statement asserts the existence of something, so the best way to demonstrate that it is true is simply to exhibit such an object.

<u>Consider the number 1.</u>　　<u>Since $1^3 = 1$, the statement is true.</u>

Produce a candidate.　　Show that it does what you want.

We usually have to work harder than this to produce the object that we need, but the process of proving an existence theorem is always the same. Suppose we want to prove that "There exists a clacking waggler." We prove existence theorems in two steps.

1. We produce a "candidate." That is, we describe an object that we claim should be a clacking waggler.
2. We show that our candidate actually is what we claim it is. In this case, we show that it is a waggler and that it clacks.

> One thing about existence proofs may seem baffling at first. When a candidate is produced, the proof need not tell you where the candidate came from or why it was chosen, just as a chess player need not tell you what strategy she or he used to decide what move to make. The only mathematical requirement is that the candidate be given explicitly and that the proof show the candidate does what it is meant to do.

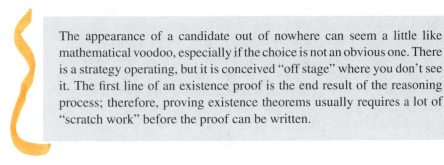

The appearance of a candidate out of nowhere can seem a little like mathematical voodoo, especially if the choice is not an obvious one. There is a strategy operating, but it is conceived "off stage" where you don't see it. The first line of an existence proof is the end result of the reasoning process; therefore, proving existence theorems usually requires a lot of "scratch work" before the proof can be written.

Consider now the statement

There exists a line in the plane passing through the points $(-1, 1)$, $(2, -1)$, and $(3, 0)$.

I presume that in the "thought experiment" you decided this statement was false. For good measure, we will think through how to prove this, using some of the language we have been developing.

Saying that the statement is false is the same as saying that its negation is true. In Exercise 1.8.5 you showed that the negation of this statement is (something like)

If ℓ is a straight line, ℓ fails to pass through at least one of the points $(-1, 1)$, $(2, -1)$, and $(3, 0)$.

So in order to prove that the statement is false, we have to examine *every straight line in the plane* and show that at least one of the three points fails to lie on it. Starting from the notion that straight lines are of the form $\ell(x) = mx + b$, an argument might go something like this.

Proof.   We start with the straight line $\ell(x) = mx + b$. If the points $(-1, 1)$ and $(2, -1)$ are to lie on the line, the following must be true:

$$1 = m(-1) + b \quad \text{and} \quad -1 = m(2) + b.$$

Solving these two equations simultaneously, we see that the only possibilities for m and b are $m = -2/3$ and $b = 1/3$. Therefore, the only line that has any chance of containing all three points is $\ell(x) = (-2/3)x + 1/3$. But it is not true that $0 = (-2/3)(3) + 1/3$, so $(3, 0)$ does not lie on the line. The three points do not all lie on a single line.   ■

1.10   Uniqueness Theorems

Many mathematical objects are unique. That is, there is only one of them: cube roots of real numbers, inverses of functions, solutions to differential equations (under suitable conditions).[2] You will occasionally be called upon to prove the uniqueness of a mathematical object (frequently right after you have proved its existence). Suppose you know that there is a clacking waggler, and you want to show that it is unique. That is, that there is only one clacking waggler. You do this by assuming that you have two clacking wagglers and demonstrating that they must be the same. A theorem that guarantees the uniqueness of a mathematical object is called a **uniqueness theorem**.

> Contrary to popular usage, the word *unique* does not mean "distinctive" or "idiosyncratic," it means (literally) "one of a kind." Thus, if we say that some object is unique, we mean that there is only one.

1.10.1 EXAMPLE

Assume that $x^3 + 37$ has a real root. (This is true. All polynomials of odd degree have at least one real root.) Prove that it has only one.

Proof.   Assume that x_1 and x_2 are real numbers and that $x_1^3 - 37 = 0$ and $x_2^3 - 37 = 0$. Then $x_1^3 - 37 = x_2^3 - 37$. So $x_1^3 = x_2^3$. Since cube roots of real numbers are unique, $x_1 = x_2$. ∎

I actually proved a stronger uniqueness result. Can you see what it is?

1.11   Examples and Counterexamples

In Section 1.5 we said that in order to prove that an implication is false, we need only provide a counterexample. That is, if A and B are predicates, the statement "If A, then B" is true, unless there is some value of the variable(s) that makes A true and B false. When we provide a counterexample, we are just showing that such a value exists.[3]

1.11.1 EXERCISE

Give counterexamples to the following proposed (but false) statements.

1. If a real number is greater than 5, then it is less than 10.
2. If x is a real number, $x^3 = x$.
3. All prime numbers are odd numbers. *What is the hypothesis here? What is the conclusion?*

---

[2] Never underestimate a theorem that tells you that if you have one you have them all. Uniqueness theorems are very powerful. You probably use some uniqueness theorems by reflex without even thinking about them. You will run into them a lot as you continue your mathematical studies.

[3] In effect, providing a counterexample is sort of an existence proof. And it's handled pretty much the same way. Give the example and show that it makes the hypothesis true and the conclusion false.

4. If $x + y$ is odd and $y + z$ is odd, then $x + z$ is odd.

5. Given three distinct points in space, there is one and only one plane passing through all three points. □

1.11.2 EXERCISE

Think again about the implications given in the "thought experiment." When you decided that one of them was false, did you justify your conclusion by means of a counterexample? If there are some you haven't justified satisfactorily, does this language help you to "fill out" the arguments? □

Counterexamples give us a straightforward procedure for showing that an implication is false. But how do we prove that an implication is true? Before we answer this question, I want to discuss a final preliminary issue. Let's consider the statement

Every positive integer is the sum of distinct powers of two.

In trying to evaluate the truth or falsehood of a theorem like this, I start by trying a lot of examples.

- $3 = 2 + 1 = 2 + 2^0$.
- $5 = 4 + 1 = 2^2 + 2^0$.
- $6 = 4 + 2 = 2^2 + 2$.
- $7 = 2^2 + 2 + 2^0$.
- $9 = 2^3 + 2^0$.

(So far, so good. Let's skip around a bit.)

- $29 = 2^4 + 2^3 + 2 + 1$.
- $113 = 2^6 + 2^5 + 2^4 + 2^0$.

(OK, so what about *really big* numbers?)

- $5,678,984 = 2^{22} + 2^{20} + 2^{18} + 2^{17} + 2^{15} + 2^{13} + 2^{10} + 2^9 + 2^8 + 2^7 + 2^3$.

All these still wouldn't quite convince me, so I would try a bunch more. As I check more and more cases, I begin to think that the statement is probably true.

The hunch that makes me want to draw this conclusion is called **inductive reasoning.** It is the process by which we draw conclusions about the general based on the particular. (That is, we look at some examples, identify a common element, and then guess that the common element holds in all cases.) This is contrasted with **deductive reasoning,** which is the process of using the rules of logic to deduce logical consequences from assumed premises or previously proved theorems. Things can only be conclusively proved by deductive reasoning. In the preceding example, checking 100 or 1000 cases might strengthen my hunch, but it still would not prove anything conclusively.

Inductive reasoning is still important for mathematicians, of course, because it is the tool by which we make conjectures. Once you think you know what is true, you can concentrate on finding a proof. But making a guess about what is true—even a very informed guess—is simply not the same as proving it.

I'm not just being picky here. To show you how dangerous it is to assert that something is proved based on having checked even a large number of examples, consider the following example found in *Induction in Geometry* by L. I. Golovina and I. M. Yaglom.

1.11.3   EXAMPLE

Consider the polynomial $991n^2 + 1$. Suppose that you were to start evaluating this polynomial at successive positive integers at the rate of one per second. You would never get a perfect square. Not because it never *is* a perfect square, but because it would take you on the order of 4×10^{20} years to find the smallest natural number n for which it is. (The age of the universe is about 1.5×10^{10} years.) The smallest natural number n for which $991n^2 + 1$ is a perfect square is:

$$n = 12{,}055{,}735{,}790{,}331{,}359{,}447{,}442{,}538{,}767 \approx 1.2 \times 10^{28}.$$

(Check, it works.)   ∎

Moral.   Providing a counterexample is **conclusive proof** that an implication is false, but checking even a large number of examples (unless you can exhaust all possible cases!) doesn't prove an implication in general.

1.12   Direct Proof

Remember that if A and B are predicates, the statement "If A, then B" is true if for all values of the variable(s) that make A true, B is true, also. One way to prove "If A, then B" is to check all possible cases where the hypothesis holds and see if the conclusion is also true. Of course this becomes cumbersome if the number of cases is large and impossible if it is infinite. We certainly cannot check them one by one. So we assume an abstract situation in which the hypothesis holds (nothing can be assumed beyond the hypothesis itself) and show that the conclusion must hold also.

Thinking about an example should help. Consider the statement "If x is an even integer, then x^2 is an even integer." I suspect that when you conducted the "thought experiment" you decided that this is true. It is a case in which there are infinitely many values of x that make the hypothesis true. So we will have to assume (in the abstract) that x is even and then show that x^2 has to be even, too.

If we are to get anywhere, we first have to recall what it means to say that an integer is even:

When an implication is proved by assuming that the hypothesis is true and then showing that the conclusion is also, the proof is called a **direct proof**.

Let z be an integer. Then z is said to be even *if there exists an integer w such that $z = 2w$.*

Here is the proof that if x is an even integer, then x^2 is an even integer.

Proof.   Suppose that x is an even integer. Then by definition of even integer, we know that there must exist an integer y such that $x = 2y$. Now we have to show that there is an integer w so that $x^2 = 2w$. Let $w = 2y^2$. Since the product of integers is an integer, $w = 2y^2$ is an integer. Notice that

$$x^2 = (2y)(2y) = 2(2y^2) = 2w.$$

Thus x^2 is an even integer.   ■

This argument works for *any* even number; thus all cases have, in some sense, been checked.

> **The Role of Definition:** The engine that drove our argument was the *definition* of even number. The vague notion that even integers are those in the list $0, \pm 2, \pm 4, \pm 6, \pm 8, \ldots$ could not give us the power we need to prove the theorem. As I stressed in the Introduction, definitions are *tools* that we use to express abstract concepts using mathematical statements. Without careful definitions, we have nothing on which to apply the rules of logic. Insight comes from an intuitive understanding of what the terms mean, from checking examples, and so forth. But theorems are *proved* by applying logical principles to abstract definitions.

1.12.1   EXERCISE

If you haven't done so already, use a direct proof to prove that "If $x + y$ is even and $y + z$ is even, then $x + z$ is even."   □

Besides the direct proof, two other methods for proving theorems are very commonly used: proof by contrapositive and proof by contradiction.

1.13   Proof by Contrapositive

1.13.1   EXERCISE

Let A and B be predicates. Construct a truth table to show that the following statements are equivalent:

$$A \Longrightarrow B \quad \text{and} \quad {\sim}B \Longrightarrow {\sim}A.$$ □

Remember, we have said that equivalent statements can be thought of as the same statement expressed in different ways. In this case, "If not B, then not A" should be viewed as a rephrasing of "If A, then B."

1.13.2 DEFINITION

The statement $\sim\!B \Longrightarrow \sim\!A$ is called the **contrapositive** of the statement $A \Longrightarrow B$.

1.13.3 EXERCISE

Find the contrapositives of the following statements. Write things in positive terms wherever possible.

1. If $x < 0$, then $x^2 > 0$.
2. If $x \neq 0$, then there exists y for which $xy = 1$.
3. If x is an even integer, then x^2 is an even integer.
4. If $x + y$ is odd and $y + z$ is odd, then $x + z$ is odd.
5. If f is a polynomial of odd degree, then f has at least one real root.   □

1.13.4 EXERCISE

Use your intuition about implication to explain why "If A, then B" and its contrapositive are saying the same thing.   □

Sometimes it is easier to prove the contrapositive of a statement than it is to prove the statement itself. (The contrapositive gives you different statements to work with which may simply be more tractable.) Since they are equivalent, proving "If not B, then not A" is the same as proving "If A, then B." A proof in which the contrapositive is proved instead of "If A, then B" is called a **proof by contraposition** or a **proof by contrapositive**.[4]

Since the contrapositive of an implication is itself an implication, the *procedure* for doing a proof by contraposition is to figure out what the contrapositive is (as in Exercise 1.13.3) and then to proceed exactly as one does in a direct proof.

1.14 Proof by Contradiction

1.14.1 EXERCISE

Let A, B, Q, and P be statements. Construct a truth table to show that the following statements are equivalent:

$$Q \quad \text{and} \quad (\sim\!Q) \Longrightarrow (P \wedge \sim\!P).$$

---

[4] The latter is less grammatical, but more commonly used!

In particular, explain why this means that:

$$A \Longrightarrow B \quad \text{and} \quad (A \wedge \sim B) \Longrightarrow (P \wedge \sim P)$$

are also equivalent. (Note that logically the statement P need have no connection what-soever with the statements A, B, or Q, though it often does in practice.) □

1.14.2 EXERCISE

To help you see why this equivalence makes sense, suppose you have statements X and Y in which

$$X := (A \Longrightarrow B), \text{ so that } \sim X := A \wedge \sim B, \text{ and}$$
$$Y := P \wedge \sim P.$$

If you know that $\sim X \Longrightarrow Y$ is true and Y is false, what can you say about the truth value of X? □

Recall that "P $\wedge \sim$P" is a contradiction, a statement that is always false. (Clearly, a statement and its negation cannot both be true.) There is a proof technique, called **proof by contradiction,** in which we first assume that the statement we want to prove is false and then show that this implies the truth of something that we know to be false. (For instance, if your reasoning ends in the conclusion that $1 = 0$, you have arrived at a contradiction since $1 = 0$ and $1 \neq 0$ cannot both be true.)

To be specific, suppose that we want to prove that $A \Longrightarrow B$ is true. We know that the negation of $A \Longrightarrow B$ is $A \wedge \sim B$. Exercise 1.14.1 tells us that if we assume $A \wedge \sim B$ and can reason our way to a contradiction, we will be able to conclude that $A \Longrightarrow B$ is true. When we do this, we are doing a **proof by contradiction.**

1.14.3 EXAMPLE (Proof by contradiction)

If $a > 0$, then $1/a > 0$. ∎

Proof. The proof is by contradiction. Thus we assume the hypothesis ($a > 0$) and the negation of the conclusion ($1/a \leq 0$). Since $1/a \leq 0$, there is some nonnegative number b so that

$$1/a + b = 0$$

Multiplying both sides by a, we get

$$1 + ab = 0.$$

Since $a > 0$ and $b \geq 0$, $ab \geq 0$. Hence $1 \leq 0$. Since we also know that $1 > 0$, we get the desired contradiction. We therefore conclude that our orginal assumption must have been false, so

If $a > 0$, then $1/a > 0$

is true. ∎

1.14.4   EXERCISE

Use proof by contradiction to prove that "If x is an integer, then x cannot be both even and odd."   □

Remark.   The problem with the preceding proofs is that each proof necessarily appealed to statements that were at least as in doubt as the statements that were being proved. (Go back and look at the proofs. Can you see where?) This is the basic problem with exercises that ask you to prove mathematical statements before any real mathematics has been discussed. They almost always amount to proving statements that we already "know" to be true using other statements that we already "know" to be true.

You should think of the proofs in this chapter as giving you practice only in using the "logical form" of the proof techniques. You should not think of them as having mathematical content. After this chapter, the proofs that you do should appeal only to the definitions and theorems that have already been discussed in a mathematically rigorous way. When you do this, you will know exactly what the starting assumptions were. You will be building a strong chain of mathematical reasoning whose beginning and end you can see. You will, thus, be standing on much firmer mathematical ground than you have been in the preceding proofs.

1.15   Proving Theorems: What Now?

In the latter part of this chapter, we have talked a bit about the logical basis for several proof techniques: direct proof, proof by contraposition, proof by contradiction. And we have talked about what general approach to take when proving existence and uniqueness theorems. But caution! You are far from being an expert on how to use these techniques; therefore, **you are not leaving this chapter!** You are just beginning to use it. In the course of working through the mathematics in the pages that lie before you, you should turn back to this chapter on a regular basis (at least for a while). When trying to decide on a strategy for proving something, review the various proof techniques and weigh them as options in your mind. Sometimes I will give you a hint as to what I would do. However, there is rarely only a single way of doing things. If my hint doesn't seem to be working for you, try something else. Proving theorems is a creative process. You may create something different from your neighbor. One proof may be shorter or more elegant or more revealing or simply more aesthetically pleasing than another, and you may want to strive for such improvements as you get more proficient. But the bottom line is that a proof is a proof is a proof. For now, concentrate on finding sound arguments that will prove the theorems you encounter. Use any and all tools at your disposal.

■ PROBLEMS

1. Suppose we understand the free variable z to refer to (a) books, (b) automobiles, and (c) pencils. For each context,

   ■ Give an example of a predicate $A(z)$ for which "For all z, $A(z)$" is a true statement.
   ■ Give an example of a predicate $B(z)$ for which "For all z, $B(z)$" is false but "There exists z such that $B(z)$" is true.

2. Is it possible to have a predicate $T(x)$ such that "For all x, $T(x)$" is true, but "There exists some x such that $T(x)$" is false? Justify your answer.

3. Consider the statements

   $P :=$ "Dogs eat meat."
   $Q :=$ "Rome is in Italy."
   $R :=$ "Chocolate prevents cavities."
   $S :=$ "The moon is made of green cheese."

   Determine whether each of the following is true or false.

   (a) If P, then Q.   **(b)** If P, then R.   **(c)** If R, then S.
   (d) If S, then Q.   **(e)** If Q, then S.

4. With apologies to Sidney Harris for trodding on his terrific cartoon (shown in Figure 1.1), I'd like to play a little with the dog's statement. Consider the assertions made by the dog:

   $A :=$ "All cats have four legs."
   $B :=$ "I have four legs."
   $C :=$ "I am a cat."

   (Are these assertions statements or predicates? Explain.)
        The dog's statement is of the form "if A and B, then C."

   (a) Construct a truth table for the statement "If A and B, then C."

   (b) Now consider the actual truth values of the assertions made by the dog. Cross out the lines of the truth table that don't apply *in this particular instance*. What do you see?

5. This problem refers to the equivalence discussed in Example 1.7.6.

   (a) Using your intuition about implication, explain why it makes sense to say that

   > *If A is true, then either B is true or C is true*

   means the same thing as

   > *If A is true and B is false, then C is true.*

   (b) Go back to the "thought experiment" in Section 1.1. Find a statement that is written in the form "If A, then B or C." Find two equivalent rephrasings of the statement. (Did you intuit these rephrasings when you worked with the problem during the thought experiment?)

   (c) Construct a truth table to show that $(A \implies (B \lor C))$ *cannot* be rephrased as $((A \implies B) \lor (A \implies C))$. Using the statement you discussed in part (b) as an example, explain why.

6. Consider the statement "Marlene has brown hair." When asked to negate this statement, some students are apt to say, "Marlene has blond hair." Explain why this is incorrect. (*Hint:* There

is an important difference between a statement that is false when "Marlene has brown hair" is true, and the negation of "Marlene has brown hair.")

(For a more mathematical, but parallel, example: Ask yourself why $x = 3$ is not the negation of $x = 7$. What *is* the negation of $x = 7$?)

7. Suppose that (a, b) and (c, d) are two distinct points in $\mathbb{R}^2$. Use the processes described in Sections 1.9 and 1.10 to prove that there exists a unique line passing through the two points. (Remember that the work you do to determine the candidate in the existence part is not part of your proof.)

8. Describe what you would have to do to show that an object is *not* unique.

9. Your goal will be to prove that "If x is an odd integer, then x^2 is an odd integer."

   (a) Here are two possible definitions for an odd integer.
      ▪ An integer z is odd if it is not even.
      ▪ An integer z is odd if there exists an integer w such that $z = 2w + 1$.
      Which of these two definitions do you think will be more useful to you in the proof? Why?

   (b) Prove that the square of an odd integer is odd.

■ QUESTIONS TO PONDER

This is the first in a series of sections titled **Questions to Ponder.** In these sections you will find questions that you can play with at your leisure; you may work with other students in your class or challenge friends that are not in your class. Some of the questions should be resolvable with a bit of work. Some will become tractable as you proceed through the book. Some will be harder, and you may not be able to solve them completely, but I will only include such problems if you can make some progress on them at least by looking at examples. Some questions are philosophical in nature and their answers may be open-ended.

1. The following two statements were given as alternate definitions for an odd integer:
   ▪ An integer z is odd if it is not even.
   ▪ An integer z is odd if there exists an integer w so that $z = 2w + 1$.

   One would hope that these definitions are equivalent. (That one is just a rephrasing of the other!) Can you prove this?

2. You should try to prove that $\sqrt{2}$ is irrational. (*Remember:* A rational number is a number that can be written as a ratio of integers. One classic proof assumes $\sqrt{2}$ is rational; that is, $\sqrt{2} = \frac{m}{n}$. What can you say about the prime factors of m^2 and $2n^2$ and what does this tell you?)

3. Try to prove that there are infinitely many prime numbers.

4. Try to prove that every positive number has a positive square root. (That is, prove that "For any positive real number x there exists a positive real number y such that $y^2 = x$.")

5. Can every positive integer be written as the sum of distinct powers of two?

6. Can every even integer greater than 2 be expressed as the sum of two prime numbers? (This is the famous "Goldbach conjecture." The question was first asked in 1742. Mathematicians continue to struggle with it today. No one knows the answer.)

2 Sets

A set is a many that allows itself to be thought of as a one.—*Cantor*

2.1 Sets and Set Notation

Mathematical structures are built from sets. We could in principle begin with the axioms of set theory, deducing from them all of the useful properties of sets and constructing such familiar mathematical objects as real numbers, functions, and so on. We are not going to do this. A trip through axiomatic set theory is time-consuming and can sometimes obscure essentially simple ideas about sets. Furthermore, most mathematicians can do their work without going all the way back to the underlying axioms of set theory. Intuitive ideas about sets and a few useful theorems (all of which are justifiable from the axioms) suffice for most purposes.

In this chapter, we will develop many of these set-theoretic ideas and theorems without a direct appeal to the axioms. We will assume without further comment the elementary arithmetic properties of the natural numbers:

$$\mathbb{N} = \{1, 2, 3, 4, \ldots\}.$$

In examples and exercises we will take advantage of the reader's informal understanding of the integers:

$$\mathbb{Z} = \{0, \pm 1, \pm 2, \pm 3, \pm 4, \ldots\},$$

the rational numbers $\mathbb{Q}$ (quotients of integers), the real numbers $\mathbb{R}$ (both rational and irrational numbers), and the Cartesian plane $\mathbb{R}^2$ (ordered pairs of real numbers).[1]

---

[1] Justifying these assumptions in retrospect, in Appendix A we set out some axiomatic set theory and build the natural numbers, and we give some indication of how to derive their fundamental arithmetic properties; divisibility and prime factorization are discussed in Chapter 6. In Chapter 8 we discuss the axioms for the real number system. In Appendix B we build the real numbers from set theory.

Though an intuitive study of sets can take us far, it is important to know that there are pitfalls. A careless use of ideas about sets can lead to paradoxes. It is for this reason that the axioms are needed in the broader mathematical context. To illustrate what we mean by this, in Section 2.6 I include a short discussion of the most famous set-theoretic paradox—Russell's paradox.

Set, element, and "∈" are the undefined terms of set theory. Intuitively, we think of a set as a collection of things. An element is one of the things that lies in the set. The symbol ∈ indicates a relationship between a set and its elements. We write "$x \in A$" if x is an element of the set A. (If $x \in A$, we may equivalently say that x is in A.) If x is not an element of the set A, we write $x \notin A$.

We may intuitively picture a set as "a bunch of things in a box." If we take those things and put them in another box, the set does not change. Furthermore, if we rearrange the things inside the box, we do not change the set. In other words, what counts is the contents of the box, and not the box itself or the arrangement of its contents. A set is characterized entirely by its elements.

Before we can say much more about set theory, we must consider notation for describing sets. To describe a set, we must say what elements are in the set. It is, of course, sufficient to list the elements of the set. For example:

- $A = \{a, b, c, d\}$.
- $S = \{\{1, 2\}, \{1, 3, 5\}, \{1, 6, 8\}\}$
  (Sets can contain other sets as elements. In this example, the number 1 is not an element of S, but the set containing the numbers 1 and 2 is!)

However, this approach becomes impractical for large sets, impossible for infinite sets, and in any case may not be very enlightening. Consider, for instance, the set $E = \{6, 12, 30\}$. Though this is a perfectly fine description of E, it might perhaps be more informative to describe it thus:

$$E = \{e \in \mathbb{N} : e \text{ is the number of edges on one of the five platonic solids}\}.$$

(There are only three numbers in this set because the cube and the octahedron both have 12 edges. The icosahedron and the dodecahedron both have 30.)

As in this example, we frequently describe the contents of a set by means of some property that is satisfied by every element of the set and not by things that are not elements of the set. Then we know that everything that satisfies the property is in the set and things that don't satisfy the property are not. Here are some more examples.

2.1.1 EXAMPLE

1. $S = \{(x, y) \in \mathbb{R}^2 : x^2 + y^2 = 16\}$.
2. $S = \{p \in \mathbb{N} : p \text{ is prime}\}$.
3. $S = \{P : P \text{ was a President of the United States}\}$.

(To help make certain that you understand the notation, give a geometric description of the first set, and list three elements of each of the other two sets.)   ∎

Reading aloud:

- $A = \{a, b, c, d\}$ is read as "A is the set containing a, b, c, and d."
- $E = \{e \in \mathbb{N} : e$ is the number of edges on one of the five platonic solids$\}$ is read as "E is the set of all e in the natural numbers such that e is the number of edges on one of the five platonic solids."
- $S = \{(x, y) \in \mathbb{R}^2 : x^2 + y^2 = 16\}$ is read as "S is the set of all pairs (x, y) in $\mathbb{R}^2$ such that $x^2 + y^2 = 16$."

These descriptions are of the form $S = \{x \in X : \mathrm{P}(x)\}$, where $\mathrm{P}(x)$ is a predicate in the free variable x and X is the set that gives the range of acceptable values for x. The set S described in this way consists of the elements of X that make $\mathrm{P}(x)$ a true statement. When there is absolutely no possibility of confusion, we may omit explicit reference to the set X, as in the third set given in Example 2.1.1. (Be sure you can identify the appropriate predicate in each of the sets described in Example 2.1.1.)

2.1.2   EXAMPLE (Interval notation)

Since we often need to talk about *intervals* in $\mathbb{R}$, we use the familiar **interval notation** to denote them.

Suppose that a and b are fixed real numbers. Then

$$[a, b] = \{x \in \mathbb{R} : a \leq x \leq b\}$$
$$[a, b) = \{x \in \mathbb{R} : a \leq x < b\}$$
$$(a, b] = \{x \in \mathbb{R} : a < x \leq b\}$$
$$(a, b) = \{x \in \mathbb{R} : a < x < b\}$$

$[a, b]$ is called the **closed interval** from a to b. (a, b) is called the **open interval** from a to b. $[a, b)$ and $(a, b]$ are referred to as **half-open intervals**.   ∎

It is convenient on many occasions to be able to refer to a set with no elements. (For instance, sets of the form $S = \{x \in X : \mathrm{P}(x)\}$ may produce such a set. If there are no $x \in X$ that make $\mathrm{P}(x)$ true, then S contains no elements. Consider $\{x \in \mathbb{N} : x^2 < 0\}$.)

2.1.3   DEFINITION

The set with no elements is called the **empty set**; it is denoted by $\emptyset$. (In some books you will find the empty set denoted by $\{\ \}$.)

2.2   Subsets

Often we deal with several sets at once and study the relationships between them. Here is one relationship that is probably already familiar to you.

2.2.1   DEFINITION

If A and S are sets, we say that S is a **subset** of A if every element of S is in A (which can be written in implication form as "If $x \in S$, then $x \in A$.") We denote this by $S \subseteq A$.

The next theorem shows that every nonempty set has at least two subsets.

2.2.2   THEOREM

For all sets X, $\emptyset \subseteq X$ and $X \subseteq X$. (*Hint:* Look at the implication given in the definition of subset. Explicitly write out the implications that you are meant to prove. What can you conclude?)   □

2.2.3   EXERCISE

How many subsets does $\emptyset$ have?   □

Notice that the statement "If $x \in S$, then $x \in X$" is an implication just like those that we studied in the last chapter. Thus to show that $S \subseteq X$, we must take an x for which the hypothesis is true (that is, an x in S) and show that the conclusion is also true for that x. That is, we must show $x \in X$. We do this so often in mathematics that there is a name for the process. It is called an **element argument**. To get started, try your hand at the following (very simple) element argument.

2.2.4   EXERCISE

Prove that if $A \subseteq B$ and $B \subseteq C$, then $A \subseteq C$. (Notice that your goal is to prove that $A \subseteq C$. Therefore, according to the process described above, you begin your argument with "Let $x \in A$." At the end of your argument, you should be able to say: "Then $x \in C$.")   □

2.2.5   DEFINITION

If B is a subset of X and $B \neq X$, then we say that B is a **proper subset** of X.

2.2.6   EXERCISE

Give two proper subsets of the set $\{1, 3, 5, 7, 9, 11\}$.   □

Because a set is completely characterized by its elements, it is reasonable to say that two sets are the same set if they have the same elements. That is, every element of A is in B and every element of B is in A.

2.2.7 DEFINITION

Suppose that A and B are sets. Then $A = B$ if $A \subseteq B$ and $B \subseteq A$.

A word about definitions: Although definitions are written in the same form as theorems, they are fundamentally different. Suppose we have the following definition: "If X is a clacking waggler and all subsets of X are marint, then X is a **supreme clacking waggler**." Because we are giving a definition, we are saying *exactly* what we mean by supreme clacking waggler. We are not just describing some possibility for supreme clacking waggler. The statements

"X is a supreme clacking waggler," and
"X is a clacking waggler and all subsets of X are marint"

mean *exactly the same thing*. Definitions are always equivalences. By convention, the if and only if is understood and never said explicitly.

2.3 Set Operations

The axioms of set theory allow us to "build new sets from old ones." They tell us that any subset of a set that we have at our disposal is also a set. They also allow us to take unions, intersections, and complements of sets.

2.3.1 DEFINITION

Let U be a set. Let $S \subseteq U$. Define

$$S_U^{\complement} = \{x \in U : x \notin S\}.$$

The set $S_U^{\complement}$ is called the **complement** of S in U. If the set U is understood, we may just call $S_U^{\complement}$ "the complement of S" and denote it by $S^{\complement}$.

(For technical reasons having to do with set-theoretic paradoxes, complements must always be taken relative to a larger set—see Section 2.6. In the absence of the set U, $S^{\complement}$ makes no sense.)

2.3.2 EXERCISE

Consider the intervals $U = [-5, 5]$ and $S = [-5, 2]$. Find $S_U^{\complement}$. ☐

2.3.3 EXERCISE

As in Exercise 2.3.2, let $S = [-5, 2]$. What is $S_{\mathbb{R}}^{\complement}$? ☐

2.3.4   DEFINITION (Unions and intersections)

Let A and B be sets.

1. $A \cup B = \{x : x \in A$ or $x \in B\}$ is called the **union** of the sets A and B.
2. $A \cap B = \{x : x \in A$ and $x \in B\}$ is called the **intersection** of the sets A and B.

2.3.5   EXERCISE

Consider the following pair of sets:

$$A = \{a, b, c, d, e, f, g\} \quad \text{and} \quad B = \{a, e, i, o, u\}.$$

Find:

(a) $A \cap B$.   (b) $A \cup B$. □

2.3.6   DEFINITION

If A and B are sets that do not intersect (that is, if $A \cap B = \emptyset$), we say that A and B are **disjoint**.

2.3.7   EXERCISE

Give an example of a pair of disjoint subsets of $\mathbb{R}$. □

2.3.8   EXERCISE

Let A, B, and X be sets. Prove that $X \subseteq A \cap B$ if and only if $X \subseteq A$ and $X \subseteq B$. (*Note:* In this problem you have to show that two statements are equivalent. See page 24 if you need a reminder about how to do this.) □

2.3.9   EXAMPLE

It is often useful to represent simple set relationships visually using **Venn diagrams.** Figure 2.1 shows two Venn diagrams depicting sets A and B that are both subsets of a larger set U. In the diagram on the left, the shaded region shows $A \cap B$. The shaded region in the diagram on the right shows $A \cup B$. ∎

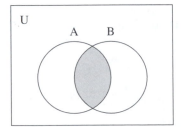 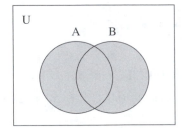

Figure 2.1   Venn diagram

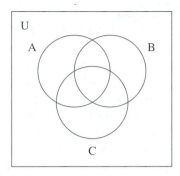

Figure 2.2

2.3.10 EXERCISE

Suppose you have three sets A, B, and C.

1. Use your intuition to devise definitions for $A \cup B \cup C$ and $A \cap B \cap C$. (Think about what similar definitions would look like for four, five, or more sets.)

2. The Venn diagram in Figure 2.2 shows three sets. Make two copies of the diagram on a piece of paper. Shade $A \cup B \cup C$ on one and $A \cap B \cap C$ on the other.   □

More generally, we can take unions and intersections of any number of sets. In fact, we may work with arbitrarily large or even infinite collections of sets. This requires some notation. In particular, we need the notion of an indexing set. As the use of the word index implies, an indexing set is a set that we use to keep track of a collection of sets. We introduce indexing sets by means of the next example. (We will content ourselves with an intuitive understanding of indexing sets; I do not give a rigorous definition.)

2.3.11 EXAMPLE (Indexing sets)

1. Suppose that we wish to consider the following set of intervals:

$$[0, 1], \ [0, 1/2], \ [0, 1/3], \ [0, 1/4], \ [0, 1/5], \ldots$$

There is a natural way to "index" these for easy reference.

$$I_1 = [0, 1], \ I_2 = [0, 1/2], \ I_3 = [0, 1/3], \ I_4 = [0, 1/4], \ldots.$$

In this case the sets we are working with are **indexed** by $\mathbb{N} = \{1, 2, 3, 4, \ldots\}$, the set of natural numbers. There is a first set, a second set, and so forth.

2. Consider the collection of circles of radius 1 in the plane whose center is on the x-axis. If we want to refer to this set of sets, we might identify each circle by its center. There is one circle for each point $(t, 0)$ on the x-axis. In other words, for each $t \in \mathbb{R}$ we might write

$$C_t = \{(x, y) \in \mathbb{R}^2 : x \text{ and } y \text{ satisfy the equation } (x - t)^2 + y^2 = 1\}.$$

Because each real number refers us to one of the circles and every circle is referenced by some number, the real numbers are an indexing set for the collection of circles. ∎

Any set, finite or infinite, can be an indexing set. In practice, of course, the indexing set is often related in some way to the contents of the set (as in the preceding examples). In the abstract, we usually use capital Greek letters to refer to indexing sets and small Greek letters to refer to the elements of these sets. For example, if we have a collection of sets indexed by the set Λ (lambda) we will have one set for each element α (alpha) in the set Λ. The following example will model some useful notation and language.

2.3.12 EXAMPLE

1. If Λ is some arbitrary set, and we have a collection of sets that is indexed by Λ we can compactly write this collection of sets this way:

$$\{U_\alpha\}_{\alpha \in \Lambda},$$

which we read as "the set of U-alpha's over alpha in lambda." If x is in every one of these sets, we would say:

$$x \in U_\alpha \quad \text{for all } \alpha \in \Lambda.$$

2. By convention, the Roman alphabet is usually reserved for real indexing sets. In the specific case of the set of circles described in Example 2.3.11, the corresponding language would be

$$\{C_t\}_{t \in \mathbb{R}}.$$

If x lies on at least one of the circles,

$$x \in C_t \quad \text{for some } t \in \mathbb{R}.$$

3. If the collection of sets is finite (except in rare circumstances) we index them by the set $\{1, 2, 3, \ldots k\}$ in the usual way,

$$\{A_1, A_2, \ldots, A_k\}$$

and we often use similar (but more compact) notation to denote this collection.[2]

$$\{A_n\}_{n=1}^{k} \text{ is the same as } \{A_n\}_{n \in \{1,2,3,\ldots k\}}.$$

4. If the collection is indexed by the natural numbers, there is a similar convention. $\{I_1, I_2, \ldots\}$ can be written in one of two ways:

$$\{I_n\}_{n \in \mathbb{N}} \quad \text{or} \quad \{I_n\}_{n=1}^{\infty}.$$ ∎

---

[2] You may find the notation reminiscent of the sigma-notation for writing sums, e.g., $\sum_{n=1}^{k} n^2$.

2.3.13 DEFINITION (Unions and intersections, revisited)

Suppose we have a collection of sets $\{B_\alpha\}_{\alpha \in \Lambda}$.

1. The union of all the sets is denoted by $\bigcup_{\alpha \in \Lambda} B_\alpha$, which we read as "the union over alpha in lambda of the B-alpha's." An element x is in $\bigcup_{\alpha \in \Lambda} B_\alpha$ if $x \in B_\alpha$ for some $\alpha \in \Lambda$. That is,

$$\bigcup_{\alpha \in \Lambda} B_\alpha = \{x : x \in B_\alpha \text{ for some } \alpha \in \Lambda\}.$$

2. The intersection of all the sets is $\bigcap_{\alpha \in \Lambda} B_\alpha$, which we read as "the intersection over alpha in lambda of the B-alpha's." An element x is in $\bigcap_{\alpha \in \Lambda} B_\alpha$ if $x \in B_\alpha$ for all $\alpha \in \Lambda$. That is,

$$\bigcap_{\alpha \in \Lambda} B_\alpha = \{x : x \in B_\alpha \text{ for all } \alpha \in \Lambda\}.$$

2.3.14 EXERCISE

Let $\Lambda = \{1, 2, 3\}$. Do the definitions of union and intersection given in Definition 2.3.13 correspond to the one that you gave in Exercise 2.3.10? What if $\Lambda = \{1, 2, 3, 4\}$? □

2.3.15 EXERCISE

1. Let $\{I_n\}_{n \in \mathbb{N}}$ be the collection of intervals described in Example 2.3.11.

(a) Find $\bigcup_{n \in \mathbb{N}} I_n$.   (b) Find $\bigcap_{n \in \mathbb{N}} I_n$.

How would your answer be different if the intervals were open intervals instead of closed intervals?

2. Let $\{C_t\}_{t \in \mathbb{R}}$ be the collection of circles described in Example 2.3.11.

(a) Find $\bigcup_{t \in \mathbb{R}} C_t$.   (b) Find $\bigcap_{t \in \mathbb{R}} C_t$. □

2.4 The Algebra of Sets

In this section you will be asked to prove some important set-theoretic identities. To prove each identity you will need to prove the equality of two sets. Recall that two sets are equal if each is a subset of the other. Suppose that we wish to prove that $X = Y$; then we must prove that $X \subseteq Y$ and that $Y \subseteq X$. Each of these will require an *element argument* (as described on page 42).

2.4.1 EXERCISE

Soon you will be asked to show that union distributes over intersection, and that intersection distributes over union. Drawing Venn diagrams can help you to understand what these identities are saying. So draw Venn diagrams that represent each of the following four sets.

1. $A \cup (B \cap C)$

2. $(A \cup B) \cap (A \cup C)$

3. $A \cap (B \cup C)$

4. $(A \cap B) \cup (A \cap C)$ □

2.4.2   THEOREM (Union distributes over intersection)

Let A, B, and C be sets. Then

$$A \cup (B \cap C) = (A \cup B) \cap (A \cup C).$$

Proof.   To show that $A \cup (B \cap C) = (A \cup B) \cap (A \cup C)$ we must show two things:

$$(\subseteq)\quad A \cup (B \cap C) \subseteq (A \cup B) \cap (A \cup C),$$
$$(\supseteq)\quad (A \cup B) \cap (A \cup C) \subseteq A \cup (B \cap C).$$

($\subseteq$) To prove the first inclusion, let $x \in A \cup (B \cap C)$. This means that $x \in A$ or $x \in (B \cap C)$. If $x \in A$, then x is in the union of A with any other set. So $x \in A \cup B$, and $x \in A \cup C$. Likewise, if $x \in (B \cap C)$, then $x \in B$ and $x \in C$. We can deduce from this that $x \in A \cup B$ and that $x \in A \cup C$. In either case, x is seen to be in $(A \cup B) \cap (A \cup C)$. This concludes the proof that $A \cup (B \cap C) \subseteq (A \cup B) \cap (A \cup C)$.

($\supseteq$) Conversely, to prove the second inclusion, suppose that $x \in (A \cup B) \cap (A \cup C)$. That is, $x \in (A \cup B)$ and $x \in (A \cup C)$. We know in general, of course, that either $x \in A$ or $x \notin A$. If $x \in A$, then certainly $x \in A \cup (B \cap C)$. So suppose that $x \notin A$. In this case, x must be in B because $x \in A \cup B$. Also we see that $x \in C$ since $x \in A \cup C$. We then conclude that $x \in B \cap C$. Hence here, too, $x \in A \cup (B \cap C)$. We have now shown that $(A \cup B) \cap (A \cup C) \subseteq A \cup (B \cap C)$.

Therefore, $A \cup (B \cap C) = (A \cup B) \cap (A \cup C)$.   ■

2.4.3   EXERCISE

In each part of the preceding proof, the argument is divided into two cases. In the proof of the first inclusion, the two cases were $x \in A$ and $x \in B \cap C$. In the proof of the second inclusion, the two cases were $x \in A$ and $x \notin A$. Think carefully about (and then explain) why the second argument did not split into the two cases $x \in A \cup B$ and $x \in A \cup C$, which might seem the more natural choice. (By the way, in the proof that intersection distributes over union, you will not find this little "wrinkle.")   □

Try your hand at the analogous:

2.4.4   EXERCISE (Intersection Distributes over Union)

Let A, B, and C be sets. Prove that

$$A \cap (B \cup C) = (A \cap B) \cup (A \cap C).$$ □

These two identities are special cases of:

2.4.5   THEOREM

Let Λ be an arbitrary indexing set for a collection of sets $\{B_\alpha\}_{\alpha \in \Lambda}$. Let C be a set. Then

1. $C \cup (\bigcap_{\alpha \in \Lambda} B_\alpha) = \bigcap_{\alpha \in \Lambda} (C \cup B_\alpha)$.

2. $C \cap (\bigcup_{\alpha \in \Lambda} B_\alpha) = \bigcup_{\alpha \in \Lambda} (C \cap B_\alpha)$.

That is, intersection distributes over union and union distributes over intersection.

(*Hint:* The stumbling block here is often just language. See Example 2.3.12 for some ideas.)   □

It is also sometimes useful to know that intersection distributes over intersection and union over union.

2.4.6   THEOREM

Let Λ be an arbitrary indexing set for a collection of sets $\{B_\alpha\}_{\alpha \in \Lambda}$. Let C be a set. Then

1. $C \cap (\bigcap_{\alpha \in \Lambda} B_\alpha) = \bigcap_{\alpha \in \Lambda} (C \cap B_\alpha)$.

2. $C \cup (\bigcup_{\alpha \in \Lambda} B_\alpha) = \bigcup_{\alpha \in \Lambda} (C \cup B_\alpha)$.   □

2.4.7   EXERCISE

Does complementation distribute over union and intersection? Well, not exactly. To get a sense for what is true, try drawing Venn diagrams to represent each of the following four sets.

1. $(A \cup B \cup C)^\complement$
2. $(A \cap B \cap C)^\complement$
3. $A^\complement \cup B^\complement \cup C^\complement$
4. $A^\complement \cap B^\complement \cap C^\complement$ □

2.4.8   PROBLEM

Let A and B be subsets of a set U. Prove that

1. $(A \cup B)^\complement = A^\complement \cap B^\complement$.
2. $(A \cap B)^\complement = A^\complement \cup B^\complement$.   □

The theorems proved in the previous problem are a special case of:

2.4.9   THEOREM   (De Morgan's Laws)

Let Λ be an arbitrary indexing set for a collection $\{A_\alpha\}_{\alpha \in \Lambda}$ of subsets of a set U. Then

1. $\left(\displaystyle\bigcup_{\alpha\in\Lambda} A_\alpha\right)^{\mathcal{C}} = \displaystyle\bigcap_{\alpha\in\Lambda} A_\alpha^{\mathcal{C}}.$

2. $\left(\displaystyle\bigcap_{\alpha\in\Lambda} A_\alpha\right)^{\mathcal{C}} = \displaystyle\bigcup_{\alpha\in\Lambda} A_\alpha^{\mathcal{C}}.$ □

2.4.10   DEFINITION

Let A and B be sets. The set

$$A \setminus B = \{x : x \in A \text{ but } x \notin B\}$$

is called the **set difference** of A and B. (Notice that if both A and B are subsets of a set U, $A \setminus B = A \cap B^{\mathcal{C}}$.)

Here are some (set) algebraic identities associated with set difference.

2.4.11   THEOREM

For any sets A, B, and C:

1. $C \setminus (A \cup B) = (C \setminus A) \cap (C \setminus B)$.
2. $C \setminus (A \cap B) = (C \setminus A) \cup (C \setminus B)$.
3. $B \setminus (B \setminus A) = A \cap B$.
4. $(A \setminus B) \cup (B \setminus A) = (A \cup B) \setminus (A \cap B)$.

(These identities can be proved in at least two different ways:

- set theoretically by using element arguments, and
- algebraically by converting the set differences to intersections and using the various algebraic identities that have already been estblished.

Try your hand at both.)   □

2.4.12   DEFINITION

The set described in part 4 of Theorem 2.4.11 is called the **symmetric difference** of A and B. It is often denoted by $A \triangle B$.

2.5   The Power Set

You will see that we often need to consider sets whose elements are themselves sets. Power sets are among the most important examples of such sets.

2.5.1   DEFINITION

If A is a set, then the **power set of** A is the set of all subsets of A. It is denoted by $\mathcal{P}(A)$.

2.5.2 EXERCISE

Find $\mathcal{P}(\{1\})$, $\mathcal{P}(\{1, 2\})$, and $\mathcal{P}(\{1, 2, 3\})$. ☐

2.5.3 EXERCISE

Let S be any set with n elements (where n is some natural number).

1. Looking over your answers to Exercise 2.5.2, guess how many elements $\mathcal{P}(S)$ will have.

2. Does your guess work for $\{a, b, c, d\}$ and $\emptyset$? (If not, revise your guess.) ☐

2.5.4 THEOREM

Let A and B be sets. Then $A \subseteq B$ iff $\mathcal{P}(A) \subseteq \mathcal{P}(B)$. ☐

2.5.5 THEOREM

Let A and B be sets. Then:

1. $\mathcal{P}(A \cap B) = \mathcal{P}(A) \cap \mathcal{P}(B)$, and
2. $\mathcal{P}(A) \cup \mathcal{P}(B) \subseteq \mathcal{P}(A \cup B)$. ☐

2.5.6 EXERCISE

Let A and B be sets.

1. Provide a counterexample to show that it is not necessarily true that

$$\mathcal{P}(A \cup B) = \mathcal{P}(A) \cup \mathcal{P}(B).$$

2. Is it *ever* true that

$$\mathcal{P}(A \cup B) = \mathcal{P}(A) \cup \mathcal{P}(B)?$$

That is, can you find sets A and B for which equality holds, or are $\mathcal{P}(A \cup B)$ and $\mathcal{P}(A) \cup \mathcal{P}(B)$ always different sets? ☐

2.5.7 PROBLEM

The purpose of this problem is to show that if we add one element to a finite set, we double the size of the power set. (Throughout this problem you should be thinking about the special cases you worked out in Exercises 2.5.2 and 2.5.3. They will give you insight into what is going on.)

Let S be any finite set and suppose $x \notin S$. Let $K = S \cup \{x\}$.

1. Prove that $\mathcal{P}(K)$ is the disjoint union of $\mathcal{P}(S)$ and

$$X = \{T \subseteq K : x \in T\}.$$

(That is, show that $\mathcal{P}(K) = \mathcal{P}(S) \cup X$ and that $\mathcal{P}(S) \cap X = \emptyset$.)

2. Prove that every element of X is the union of a subset of S with $\{x\}$, and that if you take different subsets of S you get different elements of X. Argue that, therefore, X has the same number of elements as $\mathcal{P}(S)$.

3. Argue that the previous two parts allow you to conclude that if S is a finite set, then $\mathcal{P}(K)$ has twice as many elements as $\mathcal{P}(S)$.

What insight does this add to the conjecture you made in Exercise 2.5.3?   □

2.6 Russell's Paradox

The mere fact that we can describe a bunch of things does not guarantee that the object we have described is a set! For instance, we may be able to think about the collection of all possible sets, but this collection cannot itself be considered a set without leading to paradox.

To see this, consider the following riddle:

In a certain town, Kevin the barber shaves all those and only those who do not shave themselves. Who shaves the barber?

Clearly the riddle has no good answer. If Kevin shaves himself, then he cannot be shaved by the barber (i.e., he cannot shave himself). We conclude then that Kevin does not shave himself. But the barber shaves all those who do not shave themselves, so Kevin must be shaved by the barber (i.e., he must shave himself). This is a colloquial rendition of a set-theoretic paradox universally called Russell's paradox after the mathematician Betrand Russell, who formulated it.

Before Russell, a number of mathematicians had used the concept of "the set of all sets." (Notice that this is just $\mathcal{S} = \{S : S \text{ is a set}\}$. It is easily described in the form $\{x : P(x)\}$ that we talked about in Section 2.1.) Note the fact that $\mathcal{S}$, being a "set" in its own right, must be an element of itself. Clearly then, there are two mutually exclusive kinds of sets: those that are elements of themselves and those that are not. We can also see that any possible set must belong to one category or the other. Consider the following subsets of $\mathcal{S}$. C is the subset of all sets that are elements of themselves, and D is the subset of all sets that are not elements of themselves. Consider the following question:

Does D belong to C or to D?

2.6.1 PROBLEM

Explain why this question leads to a paradox.   □

The contradiction that arose as a result of Russell's paradox and others like it showed the mathematical community that the concept of "set" could not be applied carelessly to any collection that could be comprehended by the human mind. To use Cantor's language, some groups of *many* do not allow themselves to be thought of as a single *one*. It became apparent that some collections could safely be called sets—the empty collection, all finite collections, and the collection of natural numbers were proposed as likely "safe" sets—and some, such as the collection of all sets, were not to be considered sets. How then was the mathematical community to tell which were safe collections and which would lead to contradiction? The answer lay in carefully considered axioms that would give criteria for identifying collections that are sets.

In addition to the empty set, finite collections, and the collection of natural numbers, the axioms tell us that any collections that can be built from sets by taking unions, intersections, subsets, and power sets (provided that these operations are handled a bit carefully) are to be considered sets. For more details see Appendix A.

■ PROBLEMS

1. Is there a distinction between $\emptyset$ and $\{\emptyset\}$? Explain.

2. Consider the following pairs of sets.

$$A = \{x \in \mathbb{N} : x \geq 12\} \qquad \text{and} \qquad B = \{x \in \mathbb{N} : x \leq 14\}$$
$$A = \{x \in \mathbb{N} : x \geq 32\} \qquad \text{and} \qquad B = \{x \in \mathbb{N} : x < 32\}$$
$$A = \{x \in \mathbb{R} : -3 < x \leq 3\} \qquad \text{and} \qquad B = \{x \in \mathbb{R} : x < 10\}$$

For each pair find:

(a) $A \cap B$.   (b) $A \cup B$.

3. **Playing with Venn diagrams.** Figure 2.2 shows A, B, and C in "general position." That means that we are assuming maximum possible overlap between the sets. (Specifically, A and B intersect, A and C intersect, and B and C intersect. Furthermore, there is a place where all three intersect.) In the absence of special information about the composition of the sets, we always draw Venn diagrams in general position, but if we know more about the situation, we incorporate that information into the diagram. The Venn diagram in Figure 2.3 shows three sets A, B, and C in which A and B intersect, A and C intersect, and B and C intersect, but

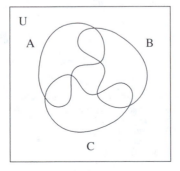

Figure 2.3

$A \cap B \cap C = \emptyset$. Draw a Venn diagram that depicts sets A, B, and C in the following special configurations.

(a) A and B are disjoint, but each has nonempty intersection with C.

(b) A is disjont from B and C, but $B \cap C \neq \emptyset$.

(c) No two of the sets intersect. (Such a collection of sets is said to be **pairwise disjoint**.)

4. (a) For each $n \in \mathbb{N}$ let

$$A_n = \left(\frac{1}{2}, \frac{1}{2} + \frac{1}{n} \right).$$

(i) Find $\bigcup_{n \in \mathbb{N}} A_n$.   **(ii)** Find $\bigcap_{n \in \mathbb{N}} A_n$.

How would your answer change if the intervals were closed instead of open?

(b) For each $r \in \mathbb{Q}$, let

$$D_r = \left(\frac{1}{2}, \frac{1}{2} + r \right).$$

(i) Find $\bigcup_{r \in \mathbb{Q}} D_r$.   **(ii)** Find $\bigcap_{r \in \mathbb{Q}} D_r$.

(c) For each $r \in \mathbb{Q}$, let

$$D_r = \left(\frac{1}{2} - r, \frac{1}{2} + r \right).$$

(i) Find $\bigcup_{r \in \mathbb{Q}} D_r$.   **(ii)** Find $\bigcap_{r \in \mathbb{Q}} D_r$.

(d) For each $r \in \mathbb{Q}$, let K_r be the set containing all real numbers *except* r. That is, $K_r = \{r\}_{\mathbb{R}}^{\complement}$.

(i) Find $\bigcup_{r \in \mathbb{Q}} K_r$.   **(ii)** Find $\bigcap_{r \in \mathbb{Q}} K_r$.

5. Suppose A and B are subsets of some set U. In this problem you will prove the following statement:

$$A \cap B^{\complement} = \emptyset \text{ if and only if } A \subseteq B.$$

It is a proposition that lends itself very well to a good review of logical principles. Note that to prove it you have to prove two implications:

$\implies$ If $A \cap B^{\complement} = \emptyset$, then $A \subseteq B$, and

$\impliedby$ If $A \subseteq B$, then $A \cap B^{\complement} = \emptyset$.

(a) For each of these implications, write out explicitly what you would have to do to prove them directly, by contrapostive, and by contradition.

(b) Now consider each of these methods of proof. Choose the one that you think makes each implication most tractable. Prove the equivalence.

6. In Theorem 2.4.11 you proved that set difference distributes over union and intersection.

(a) Do union and intersection distribute over set difference? In other words, is it true that

$$A \cup (B \setminus C) = (A \cup B) \setminus (A \cup C) \quad \text{and} \quad A \cap (B \setminus C) = (A \cap B) \setminus (A \cap C)?$$

Give a proof or a counterexample for each.

(b) Does complementation distribute over set difference? In other words, is it true that

$$(A \setminus B)^{\complement} = A^{\complement} \setminus B^{\complement}?$$

Give a proof or a counterexample.

7. Consider the symmetric difference of two sets A and B.

$$A \triangle B = (A \setminus B) \cup (B \setminus A).$$

(a) Draw a venn diagram that shows this set.

(b) We have discussed a variety of set operations. Formulate and either prove or disprove at least three distribution principles (as in the previous problem) involving symmetric difference.

8. Let A and B be subsets of a set U.

(a) Prove that $\mathcal{P}(B_U^{\complement}) \neq \mathcal{P}(B)_{\mathcal{P}(U)}^{\complement}$.

(b) Prove that $\mathcal{P}(A \setminus B) \neq \mathcal{P}(A) \setminus \mathcal{P}(B)$.

(c) Prove that $\mathcal{P}(B_U^{\complement}) \setminus \{\emptyset\} \subseteq \mathcal{P}(B)_{\mathcal{P}(U)}^{\complement}$.

(Hint: look at examples.)

■ QUESTIONS TO PONDER

1. In Problem 2.5.7, you needed to do some counting to come to the necessary conclusions. I used the word "argue" (rather than prove) in two parts of that problem because we do not, as yet, have any mathematical theory of counting. In other words, you have no definitions, axioms, theorems, or any other basis for proving your conclusions. What might such a theory of counting look like? If I say, "These two sets have the same number of elements," what does that mean mathematically? If I say, "This set has twice as many elements as that set," what does *that* mean? How might we define this rigorously?

2. Following up on this theme, suppose that two sets A and B have the same number of elements. Must their power sets have the same number of elements? Conversely, suppose $\mathcal{P}(A)$ has the same number of elements as $\mathcal{P}(B)$, does it follow that A and B have the same number of elements?

3. I did not define indexing set precisely. Can you figure out how to define indexing set?

4. Do you see a connection between Question 1 and Question 3? Does either question shed light on the other?

5. **The Berry Paradox**—*a paradox of naming.* Here is another paradox for you to ponder and try to resolve. Suppose we wish to consider "the least integer not nameable in fewer than 19 syllables." Paradox: This description has 18 syllables!

6. I recommend Appendices A and B as entire chapters full of set-theoretic questions to ponder. The Appendices provide an axiomatic treatment of set theory. The second appendix culminates in the construction of the real numbers from elementary sets. (By the middle of the second appendix, some of the gaps left to the reader are fairly big. However, following the general pattern of the ideas is not too hard and the treatment should give you a good idea of how it is all done.)

3 Induction

3.1 Mathematical Induction

Recall Exercise 2.5.3 and Problem 2.5.7. I hope that in Exercise 2.5.3 you guessed that

If $n \in \mathbb{N}$ and S is a set with n elements, then $\mathcal{P}(S)$ has 2^n elements.

I suspect you noticed that I never asked you to prove this fact, even after Problem 2.5.7 made a proof seem tantalizingly close. To see why, let's attempt to write the "obvious" proof—which, as we shall soon see, doesn't quite do the job.

Here's what happens.

"Proof." Let $S = \{a_1\}$ be any set with one element. Then $\mathcal{P}(S) = \{\{a_1\}, \emptyset\}$, which has two elements.

Let $S_2 = \{a_1, a_2\}$ be any set with two elements. Then S_2 has one more element than $\{a_1\}$. So $\mathcal{P}(S_2)$ has twice as many elements as the power set of $\{a_1\}$ by Problem 2.5.7. Since we have showed that any one element set has a power set with two elements, $\mathcal{P}(S_2)$ has $2(2) = 2^2$ elements.

Let $S_3 = \{a_1, a_2, a_3\}$ be any set with three elements. Then S_3 has one more element than $\{a_1, a_2\}$. So $\mathcal{P}(S_3)$ has twice as many elements as the power set of $\{a_1, a_2\}$. Since we have showed that any two-element set has a power set with 2^2 elements, $\mathcal{P}(S_3)$ has $2(2^2) = 2^3$ elements.

Let $S_4 = \{a_1, a_2, a_3, a_4\}$ be any set with four elements. Then S_3 has one more element than $\{a_1, a_2, a_3\}$. So $\mathcal{P}(S_4)$ has twice as many elements as the power set of $\{a_1, a_2, a_3\}$. Since we have showed that any three-element set has a power set with 2^3 elements, $\mathcal{P}(S_4)$ has $2(2^3) = 2^4$ elements.

And so on . . . □

When we finish with "and so on . . . " we are claiming that the process we have begun will continue indefinitely and that the theorem must therefore be true. Nevertheless, our proof did not explicitly address what happens with sets containing 127 elements. Have we really proved that a set with 127 elements has a power set with 2^{127} elements? This may seem like splitting hairs—after all, the phrase "and so on . . . " seems quite reasonable. Most people would probably accept the argument as a proof. But if so, what distinguishes it from the following?

"Theorem."   For all $n \in \mathbb{N}$, $n^2 + n + 41$ is a prime number.[1]

"Proof."

If $n = 1$, $n^2 + n + 41 = 43$, which is a prime number.

If $n = 2$, $n^2 + n + 41 = 47$, which is a prime number.

If $n = 3$, $n^2 + n + 41 = 53$, which is a prime number.

If $n = 4$, $n^2 + n + 41 = 61$, which is a prime number.

If $n = 5$, $n^2 + n + 41 = 71$, which is a prime number.

And so on . . .   □

An important thing distinguishes our two "theorems": the first one is true and the second one is false! For $n = 41$, it turns out that

$$40^2 + 40 + 41 = 41^2,$$

which is not prime. The phrase "and so on . . . ," which seemed okay in the first argument, is no good in the second. The second "proof" amounts to checking the first few cases and then jumping to the general conclusion that every case holds. But our argument about power sets was different. Somehow, the phrase "and so on . . . " in that situation was far more convincing.

We have to uncover the essential feature of the first argument that led us to trust the "and so on. . . ." This feature is the Principle of Mathematical Induction.

Mathematical induction is based on the following axiom of the natural numbers.

AXIOM (The Axiom of Induction)

Let S be a subset of $\mathbb{N}$ satisfying

• $1 \in S$, and
• if $k \in S$, then $k + 1 \in S$.

Then $S = \mathbb{N}$.   □

---

[1] The polynomial $n^2 + n + 41$ was first studied by the famous eighteenth-century mathematician Leonard Euler.

We are introducing a new axiom, which is a really big deal! Whenever you encounter a new axiom, you should ask yourself two questions:

1. Is this new axiom reasonable? That is, am I willing to accept it as truth without further question?

2. Is it really necessary to accept the statement as an axiom, or can I construct a proof of the statement from other already accepted principles? (Anything provable from previously accepted axioms is a theorem!)

Ideally, our axioms should be "obvious" and "intuitive." The first question calls for your judgment about how well the new axiom measures up to these criteria. The second question can lead you into deep mathematical waters pretty quickly, so you are unlikely to be able to resolve it altogether. Despite this, it is useful to give the matter some thought.

3.1.1 EXERCISE

This exercise is meant to help in your contemplation of these two important questions with regard to the axiom of induction.

1. In order to answer the first question, you must first figure out what the axiom is talking about. If you haven't already done that, do so now. Can you come to any conclusions about the first question?

2. It is instructive to try to write a proof of the Induction Axiom. Your work on the previous part of this exercise has probably given you a fair idea of how this might go. Try to write it down.   □

The "proof" that you just constructed almost certainly ended with words like "and so on. . . . " The "proof" requires infinitely many steps. In general, we do not allow proofs with infinitely many steps; for one thing, they cannot be written down completely! In this specific instance, however, we can see that it is reasonable to trust the phrase "and so on. . . . " Thus, we accept the proposition as an axiom: We think it is true, but we don't pretend that it can be proved.

Notice also that your "proof" of the Induction Axiom was reminiscent of our original argument about the size of power sets. Let's make the connection more explicit. The proposition that the power set of a set with n elements has 2^n elements is really a sequence of infinitely many statements, one for each natural number n.

- If S has 1 element, then $\mathcal{P}(S)$ has 2 elements.
- If S has 2 elements, then $\mathcal{P}(S)$ has 4 elements.
- If S has 3 elements, then $\mathcal{P}(S)$ has 8 elements.
- If S has 4 elements, then $\mathcal{P}(S)$ has 16 elements.
- *Et cetera.*

We are saved from the impossible burden of giving infinitely many proofs by the fact that the statements are *linked* in a particular way. If we can prove the first statement, and if each statement in the sequence can be shown to follow from the previous one, then there is a sort of "domino effect." The first one gives us the second, the second gives us the third, the third gives us the fourth, "and so on. . . . " If we can establish the appropriate *linkage,* then all of the dominoes will fall and we can trust the phrase "and so on. . . . " In the "rotten" argument about prime numbers, there was no linkage and therefore no domino effect.

At first, it does not seem that we have gained anything. If there are infinitely many statements in our sequence, there must be infinitely many links. However,[2] *the trick is to establish* all *of the links by proving a* single *proposition.* In the power set argument, all of the links were made by invoking the single result proved in Problem 2.5.7: Whenever we add one element to a finite set, we double the size of the power set.

We will now use the Induction Axiom to prove an important theorem that puts these ideas into a rigorous form. The Principle of Mathematical Induction, in fact, establishes a new and powerful method of proof.

3.1.2   THEOREM (Principle of Mathematical Induction)

Let P_1, P_2, P_3, . . . be a sequence of statements, one for each natural number. Suppose that

- P_1 is true, and
- if P_k is true, then P_{k+1} is true.

Then P_n is true for all $n \in \mathbb{N}$.

Proof.   In order to apply the induction axiom we must have a subset of $\mathbb{N}$. Consider the subset:

$$S = \{n \in \mathbb{N} : P_n \text{ is true}\}.$$

Because P_1 is true, $1 \in S$. Now we prove the conditional statement "If $k \in S$, then $k + 1 \in S$." Thus we suppose that $k \in S$. This means that P_k is true. By the second hypothesis in our theorem, we know that "If P_k is true, then P_{k+1} is true." Therefore, $k + 1 \in S$. The induction axiom then allows us to conclude that $S = \mathbb{N}$. That is, P_n is true for all $n \in \mathbb{N}$.   ∎

The Principle of Mathematical Induction describes a *process* that gives us our new method of proof. The process requires that we prove two statements:

1. We prove P_1 outright (this is called the **base case**), and
2. We prove an implication: "If P_k is true, then P_{k+1} is true." As always, this implication has an implied universal quantifier: We are, in fact, proving that "For all $k \in \mathbb{N}$, if P_k is true, then P_{k+1} is true." That is, the implication establishes in

---

[2] Do you have your highlighting pen ready?

one fell swoop all the linkages necessary for the domino effect. (This is called the **induction step.**)

To make the induction step, we assume that P_k is true for an arbitrary $k \in \mathbb{N}$ and show that P_{k+1} follows. The assumption that P_k is true is called the **induction hypothesis.**

Once we do these two things, the Principle of Mathematical Induction says that we are done. The beauty of all this is that we only have to accept the "and so on . . ." once—when we accept the Axiom of Induction. After that, we use the Principle of Mathematical Induction. No argument should ever end in "And so on . . ."!

In the theorem above, the base case is P_1. Then P_2 follows from P_1, P_3 follows from P_2, etc. However, it is often convenient to start with a *zeroth* statement, P_0, in which case, P_1 follows from P_0, P_2 follows from P_1, etc. Less frequently we will want to start with P_2 or even more rarely, another base case. The induction step gives us all statements after the base case. In other words, if the base case is P_{17}, we get P_{18}, P_{19}, ... (This situation can arise if the earlier cases do not make sense or they are not relevant to the theorem, or if they require a different argument than the later cases.)

3.2 Using Induction

To complete our preliminary discussion, here is an example of the way in which we implement the process of mathematical induction.

3.2.1 THEOREM

If $n \in \mathbb{N}$ and S is a set with n elements, then $\mathcal{P}(S)$ has 2^n elements.

Proof. We proceed by induction on the number of elements in the set.

Let $S = \{a_1\}$ be any set with one element. Then $\mathcal{P}(S) = \{\{a_1\}, \emptyset\}$, which has two elements.

INDUCTION HYPOTHESIS: Assume that if S is any set with k elements, then $\mathcal{P}(S)$ has 2^k elements.

Let $S = \{a_1, a_2, a_3, \ldots, a_k, a_{k+1}\}$ be any set with $k + 1$ elements. Then $S = \{a_1, a_2 \ldots a_k\} \cup \{a_{k+1}\}$. Thus problem 2.5.7 tells us that $\mathcal{P}(S)$ has twice as many elements as $\mathcal{P}(\{a_1, a_2 \ldots a_k\})$. That is, $\mathcal{P}(S)$ has $2(2^k) = 2^{k+1}$ elements.

Now the principle of mathematical induction allows us to conclude that if S is any set with n elements, its power set has 2^n elements. ∎

Here are some exercises that will allow you to practice using mathematical induction. The statements are arithmetic/algebraic in nature. You will use induction in a variety of different contexts in coming chapters. (Remember that mathematical induction works on a sequence of statements. Be sure you are very clear about what sequence of statements you are proving in each exercise.)

3.2.2 PROBLEM

For each natural number n prove that

$$\sum_{i=1}^{n} i = \frac{n(n+1)}{2}.$$

□

3.2.3 PROBLEM

Let $n \in \mathbb{N}$. Conjecture a formula for

$$a_n = \frac{1}{(1)(2)} + \frac{1}{(2)(3)} + \frac{1}{(3)(4)} + \cdots + \frac{1}{(n)(n+1)}$$

and prove your conjecture.

□

3.2.4 PROBLEM

Use induction to prove that every positive integer is either even or odd. Then use this result to show that every integer is either even or odd.

   (*Hint:* Look back at Section 1.12 for definitions of even and odd.)

□

3.2.5 PROBLEM

Let m and $n \in \mathbb{N}$. Define what it means to say that m divides n. Now prove that for all $n \in \mathbb{N}$, 6 divides $n^3 - n$.

□

3.2.6 PROBLEM

Let $x \neq 1$ be a real number. For all $n \in \mathbb{N}$,

$$\frac{x^n - 1}{x - 1} = (x^{n-1} + x^{n-2} + \cdots + x^2 + x + 1).$$

□

3.3   Complete Induction

Sometimes it is convenient to use the following (equivalent) formulation of mathematical induction. It is referred to as **complete induction**.

3.3.1 THEOREM (Principle of Complete Induction)

Let (P_n) be a sequence of statements. Suppose that

- P_1 is true, and
- If P_j is true for all $j \leq k$ then P_{k+1} is true, also.

Then P_n is true for all $n \in \mathbb{N}$. □

3.3.2 PROBLEM

Use complete induction to prove that every natural number can be written as the sum of distinct powers of two.

(*Hint:* There is a small subtlety here. To tease it out, ask yourself why the argument fails if you try to use it to prove that every natural number can be written as a sum of distinct powers of *three*—which, of course, it must.) □

Here is a very important theorem from algebra that can be proved using complete induction. Recall that a polynomial is said to be *reducible* if it can be written as the product of two polynomials of smaller degree. It is *irreducible* if it is not reducible.

3.3.3 THEOREM

Every reducible polynomial can be written as a product of irreducible polynomials.

(*Hint:* Proceed by complete induction on the degree of the polynomial.) □

3.3.4 EXAMPLE

Mathematical induction must be applied carefully. The following "proof" of an obviously false statement shows some of the possible pitfalls. Your mission is to discover how mathematical induction was misapplied.

"Theorem." All rabbits are the same color.

"Proof." Let $n \in \mathbb{N}$. Consider the statement:

Any n rabbits are the same color.

If we can show this statement is true for all $n \in \mathbb{N}$, then we will clearly have proved that all rabbits have the same color. (Since there are only finitely many rabbits.)

We proceed by complete induction on n, the number of rabbits.

Clearly, if we only have one rabbit then that rabbit will be the same color as itself.

INDUCTION HYPOTHESIS: Assume that for all $j \leq k$, any given j rabbits are the same color. We must show that any given $k + 1$ rabbits are the same color. We start by picking an arbitrary collection of $k + 1$ rabbits. Number the rabbits from 1 to $k + 1$. If we consider the 1st and the $(k + 1)$st rabbits, then these two must be the same color by the induction hypothesis. Since the 1st rabbit and the 2nd through kth rabbits are all the same color for the same reason, we must conclude that all $k + 1$ rabbits are the same color. So the principle of complete induction tells us that all rabbits are the same color. □

■ QUESTIONS TO PONDER

1. Let $m \in \mathbb{N}$ and let $S \subseteq \mathbb{N}$. Suppose that

   ▪ $m \in S$, and
   ▪ $k \in S$ implies $k + 1 \in S$.

   Prove that $\{n \in \mathbb{N} : n \geq m\} \subseteq S$. Are these sets equal? Why or why not?
       Think about how this is relevant to using mathematical induction when the base case is not 1.

2. Formulate and prove a generalization of the Principle of Mathematical Induction in which the base case is an arbitrary $m \in \mathbb{Z}$.

3. Try to prove that every nonempty subset of $\mathbb{N}$ has a least element. (This fact is very useful. It is often called the well-ordering principle.)

4. I claim in this chapter that the principles of induction and complete induction are equivalent. Can you prove it?

4 Relations

4.1 Relations

The concept of a relation is widely used in mathematics. You will find that you have already encountered it in a number of contexts. Consider, for instance, the guests at a party. Any particular guest may or may not know the name of another guest. Mathematically, we would describe this social dynamic by means of a relation. We seek a mathematical way of indicating that JoAnn knows Mark's name. This we can do with the familiar concept of an ordered pair. When we form the ordered pair (JoAnn, Mark), we understand it to mean that JoAnn knows Mark's name. (We require the order since JoAnn may know Mark's name without Mark knowing JoAnn's name.) The relation "knows the name of" is the collection of all ordered pairs (x, y), where x knows y's name. A relation is simply a set of ordered pairs.

Let A and B be sets. If $a \in A$ and $b \in B$, the ordered pair (a, b) consists of two elements in which a is designated as the first element and b is designated as the second element.[1] Two ordered pairs (a, b) and (c, d) are equal if and only if $a = c$ and $b = d$.

4.1.1 DEFINITION

If A and B are sets, the set

$$A \times B = \{(a, b) : a \in A \text{ and } b \in B\}$$

is called the **Cartesian product** of A and B. That is, $A \times B$ is the set of all ordered pairs in which the first elements come from A and the second elements come from B.

---

[1] This intuitive description of ordered pair is sufficient for our discussions. (a, b) is rigorously defined to be the set $\{\{a\}, \{a, b\}\}$. This definition not only allows us to distinguish between first and second elements, it satisfies the essential property concerning equality of ordered pairs. Can you prove using this definition that for all a and c in A and b and d in B, $(a, b) = (c, d)$ if and only if $a = c$ and $b = d$? Why isn't the definition $(a, b) = \{a, b\}$ good enough?

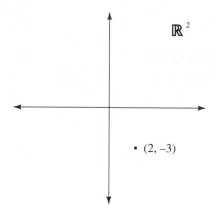

Figure 4.1   The Cartesian plane

4.1.2   EXAMPLE

The Cartesian product that you are most familiar with is probably $\mathbb{R}^2 = \mathbb{R} \times \mathbb{R}$, the Cartesian plane (see Figure 4.1).   ∎

4.1.3   EXERCISE

Let $A = \{1, 2, 3\}$, $B = \{a, b\}$.

1. Find $A \times A$ and $B \times B$.
2. Find $A \times B$.   □

4.1.4   DEFINITION

If A and B are sets, then any subset of $A \times B$ is called a **relation** between A and B. A subset of $A \times A$ is called a relation on A.

4.1.5   EXAMPLE

On the set $\mathbb{R}$, we have the relation $\leq$. The pair $(2, 3)$ is in the relation because $2 \leq 3$. The pair $(4, -\frac{1}{3})$ is not in the relation because $4 \not\leq -\frac{1}{3}$.   ∎

Remark.   The notion of a relation is a very general one. Different notation schemes are used in different contexts, and I will indicate notational conventions as we go. When talking about relations in the abstract, we will indicate that a particular pair is in the relation by some notation like $a \sim b$. (We would read this as "a is related to b.") In concrete cases, other symbols are associated with particular relations. For instance, in the case of Example 4.1.5 we write $a \leq b$ instead of $a \sim b$.

4.1.6   EXERCISE

Let S be the set of students in your class. Let B be the collection of books in your library. Define a relation between S and B.   □

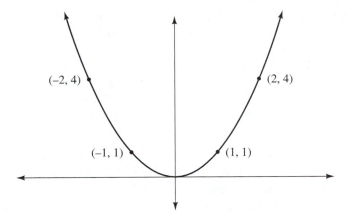

Figure 4.2 $y = x^2$

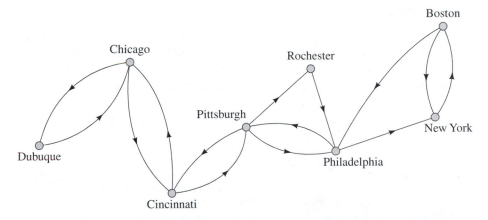

Figure 4.3 Transportation network

Many relations have useful graphical representations. Here are some examples.

4.1.7 EXAMPLE

1. Any subset of $\mathbb{R}^2$ is a relation. Figure 4.2 illustrates a familiar kind of relation. What do the pairs in this relation represent? Give three examples of pairs that belong to this relation.

2. The diagram in Figure 4.3 is an example of a *directed graph* or *digraph*. (An arrow pointing from one point to another indicates that the first is related to the second.) Speculate about what the relation might represent. ∎

For the remainder of this chapter, we will consider relations on a single set A.

We often consider relations that have convenient special properties. Reflexivity, symmetry, antisymmetry, and transitivity are some very useful ones.

4.1.8 DEFINITION

Let $\sim$ be a relation on a set A.

1. $\sim$ is said to be **reflexive** if for all $x \in A$, $x \sim x$. (That is, a relation on A is reflexive if every element of A is related to itself.)

2. $\sim$ is said to be **symmetric** if for all x, $y \in A$, $x \sim y$ implies that $y \sim x$. (A relation on A is symmetric if y is related to x whenever x is related to y.)

3. $\sim$ is said to be **antisymmetric** if for all x, $y \in A$, $x \sim y$ and $y \sim x$ imply that $x = y$. (Say in words what this means.)

4. $\sim$ is said to be **transitive** if for all x, y, and $z \in A$, $x \sim y$ and $y \sim z$ imply that $x \sim z$. (How would we say this in words?)

4.1.9 EXERCISE

Recall the relation "knows the name of" on the set of people at a party. (In this instance, instead of saying "Mark $\sim$ Karen," we would say "Mark knows Karen's name.")

Translate the definitions above into "knows the name of" language. For instance, saying that "knows the name of" is symmetric is to say that if Mark knows Karen's name, then Karen must know Mark's name. Is this true? Say in words what it would mean to say that the relation "knows the name of" is reflexive, antisymmetric, and transitive. Which of these properties holds?   □

> Note that just because we refer to things by different names does not necessarily mean that they are different; when we say $x \sim y$, it may be that $x = y$. In the case of the relations "knows the name of," any person at the party that knows his or her own name is related to him or herself.

4.1.10 PROBLEM

Indicate whether the following relations on the given sets are reflexive, symmetric, antisymmetric, or transitive. (These are not mutually exclusive conditions, so the relations may satisfy more than one.) Justify your answer.

1. $A = \{p : p \text{ is a person in Alaska}\}$. $x \sim y$ if x is at least as tall as y.
2. $A = \mathbb{N}$. $x \sim y$ if $x + y$ is even.
3. $A = \mathbb{N}$. $x \sim y$ if $x + y$ is odd.
4. $A = \mathcal{P}(\mathbb{N})$. $x \sim y$ if $x \subseteq y$.
5. $A = \mathbb{R}$. $x \sim y$ if $x = 2y$.

6. $A = \mathbb{R}$. $x \sim y$ if $x - y$ is rational. (You may assume here that the negative of a rational number is rational and the sum of two rational numbers is rational.)

· 7. $A = \{\ell : \ell$ is a line in the Cartesian plane$\}$. $x \sim y$ if x and y are parallel lines or if x and y are the same line. □

4.2 Orderings

In the previous section we mentioned the order $\leq$ on $\mathbb{R}$ as a familiar example of a relation. Order relations are fundamental to many branches of mathematics.

4.2.1 DEFINITION

A relation $\leq$ on a set A is called a **partial ordering** if $\leq$ is reflexive, antisymmetric, and transitive. A set together with a partial ordering on it is called a **partially ordered set**.[2] When we need to talk about a partially ordered set, we may refer to the *pair* $(A, \leq)$ or we may equivalently talk about the set A *under* the relation $\leq$.

Note that we use the symbol $\leq$ generically to refer to any relation that is reflexive, antisymmetric, and transitive. The underlying set may not be a set of real numbers; the symbol does not, unless specifically indicated, refer to the traditional ordering of the real numbers.

In Section 4.1 we said that we usually use the notation $a \sim b$ to indicate that the pair (a, b) is in a given relation. As the definition indicates, when that relation is a partial ordering, we customarily use $a \leq b$. In this context $\geq$, $<$, and $>$ have their usual meanings; for instance, $a > b$ means that $b \leq a$ and $b \neq a$.

4.2.2 EXAMPLE

Verify that the following are examples of partially ordered sets.

1. $\mathbb{R}$ under the traditional ordering of the real numbers.

2. For any set A, $\mathcal{P}(A)$ under $\subseteq$.

3. Any set A under the ordering: $a \leq b$ iff $a = b$. (We will call this a **totally unordered set**. Explain why this is a sensible use of language.) ∎

The second and third of these examples are different from the first in a very important way.

4.2.3 DEFINITION

A partially ordered set A with partial order $\leq$ is said to be **totally ordered** if given any two elements a and b in A, either $a \leq b$ or $b \leq a$. In this case, $\leq$ is called a **total ordering**.

---

[2] Partially ordered sets are sometimes called **posets** for short.

The definition tells us that given two elements a and b in a totally ordered set, either $a \le b$ or $b \le a$. For elements in a partially ordered set that is not totally ordered, there is a third possibility. What is it?

4.2.4   EXERCISE

Let A be a set. Show that $\mathcal{P}(A)$ need not be totally ordered under the relation $\subseteq$.   □

4.2.5   DEFINITION

Let A be a partially ordered set under $\le$. A is said to obey the **law of tricotomy** if for every a and $b \in A$ exactly one of the following is true:

$$\textbf{i.}\ a < b. \qquad \textbf{ii.}\ a = b. \qquad \textbf{iii.}\ b < a.$$

4.2.6   THEOREM

A partially ordered set A is totally ordered iff it obeys the law of trichotomy.   □

Let $(A, \le)$ be any partially ordered set and B any subset of A. Then B *inherits* the partial order from A, in the sense that if we take two elements x and y of B, we can sensibly ask whether $x \le y$, $y \le x$, or x and y are unrelated. (You should verify that the relation on B obtained in this way is a partial order.) In other words, we can easily view any subset of a partially ordered set as a partially ordered set in its own right.

4.2.7   EXERCISE

It is easy to see that $\mathcal{P}(\mathbb{N})$ is not totally ordered, but it has subsets that are totally ordered. Give an example of an infinite totally ordered subset of $\mathcal{P}(\mathbb{N})$ (under the order $\subseteq$).   □

When we are dealing with a partial order on a set with only a few elements, we can often make sense of the situation with some simple diagrams called **lattice diagrams**.[3] (Lattice diagrams can also be useful for picturing a small section of a larger ordered set.)

---

[3] One reader pointed out that this is a misnomer. Lattices are partially ordered sets in which every pair of elements has a least upper bound and a greatest lower bound. (See page 76 for definitions.) We use the diagrams to represent partially ordered sets that are not lattices. The misusage is common, so I let it stand, but the point is well taken.

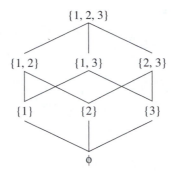

Figure 4.4   Lattice diagram for $\mathcal{P}(A)$

Consider, for instance, the power set of $A = \{1, 2, 3\}$ under set inclusion. We can draw a simple diagram that shows the order relation on this set. This is illustrated in Figure 4.4.

One element of the set is smaller than another if and only if an upward path can be found that connects them. (That path may traverse other elements in between.) Since we always move up along the diagram, elements that are on the same level cannot be related to one another, nor can an element above be less than one that sits on a level below it. Naturally, each element is related to itself as in all partial orders. This is not shown explicitly in the diagram.

4.2.8   EXERCISE

When we say that "one element of the set is smaller than another only if an upward path can be found that connects them" (even if it traverses other elements in between), what property of partial orderings are we relying on?   □

Remark.   The drawing of lattice diagrams is a pretty intuitive enterprise, so I will mostly leave it to your intuition and will not try to rigorously write out rules for constructing them. However, there is one rule that is worth mentioning explicitly. An element always appears on the *lowest possible level*. That is, an element must always be larger than at least one element on the level immediately below it. If this is not the case, we can and will draw it on a lower level.

4.2.9   EXERCISE

Show that the partial orderings on $\{a, b, c, d, e, f\}$ depicted in Figure 4.5 are the same. (Remember, partial orderings are sets of ordered pairs. Two partial orders are same if those sets are equal.) Notice, however, that only one of the three lattice diagrams is "legal." Which one is it?   □

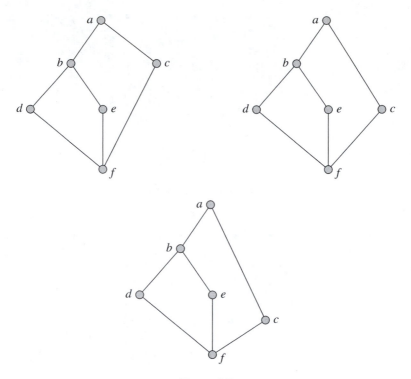

Figure 4.5

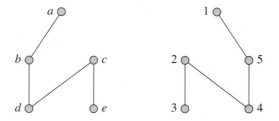

Figure 4.6   Isomorphic partial orders

Consider the diagrams shown in Figure 4.5. Suppose that we had not labeled the elements. Then we would have been observing only the order structure of the set and nothing else. There is a sense in which the two lattice diagrams shown in Figure 4.6 are the *same* mathematical structure. The difference between them amounts only to a

Figure 4.7 Partial orders on two elements

"relabeling" of the points. As long as we are interested only in the order structure, such relabeling is not important.[4]

Two mathematical structures that are the same up to a relabeling of the elements involved are called **isomorphic** structures. The partial orderings shown in Figure 4.6 are, therefore, isomorphic partial orders. We will have more to say about isomorphic partial orders when we talk about functions. For now, we will make do with an intuitive idea of what isomorphism means.

The etymology of the word "isomorphic" is enlightening. It comes from the two greek words *isos* meaning *same* and *morphos* meaning *shape*. Things that are isomorphic have the same mathematical "shape" or structure.

4.2.10 EXAMPLE

Suppose we have sets with two, three, and four elements. How might these sets be partially ordered (up to a relabeling of the elements—*up to isomorphism*)? We can use lattice diagrams to find all possible situations.

1. *Partial orders on sets with two elements.* For the set with two elements, we have only two possibilities: Either one element is smaller than the other or they are not related. This yields only two possible partial orders (which we display in Figure 4.7).

2. *Partial orders on sets with three elements.* Draw lattice diagrams for all possible partial orders on a set with three elements. There are five of them. (*Hint:* If you have trouble, you might peek at the discussion given below about partial orders on a set with four elements and then try again.)

3. *Partial orders on sets with four elements.* As the number of elements gets larger, the number of possible partial orders increases rapidly, so if we have any hope of getting all possibilities, we have to be quite systematic about the way we go about classifying them. One possible approach is to think about the number of "levels" there will be in a lattice diagram. If we have four elements, we can have our elements on one, two, three, or four levels. (If all the elements are on the

---

[4] You have seen this before. The functions
$$f(x) = x^4 - 1 \quad \text{and} \quad g(y) = y^4 - 1$$
are, in fact, the same mathematical object.

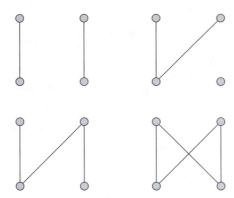

Figure 4.8   Partial orders on a set with four elements in which there are two elements on each level

same level, we have a totally unordered set. If the elements lie on four levels, we have a totally ordered set.)

Let us consider partial orders in which there are only two levels. Here there are three possibilities:

- There are two elements on each level.
- There are three elements on the first (lowest) level and one on the second level.
- There is a single element on the first level and there are three on the second level.

Consider the case in which there are two elements on each level. Remember that the rules for drawing lattice diagrams require that each element on the second level be "tied" to at least one element on the first level. This gives us four possibilities. Each element on the first level is tied to a different element on the second level. One of the elements on the first level is tied to both elements on the second level, and the other element is unrelated to the elements on the second level. One of the elements on the first level is tied to each element on the second level and the other element is tied to only one of them. The final possibility is that each element on the first level is tied to each element on the second level. This line of reasoning yields the lattice diagrams shown in Figure 4.8.

Now you should attempt to finish the classification of partial orders on sets with four elements. (There are 16 of them!)

4. Draw lattice diagrams for at least ten distinct partial orders on sets with five elements. Can you get all of them? (There are 55 in all.)   ■

Though lattice diagrams can be very useful for describing partial orderings on small sets, they are of limited usefulness when the sets we are dealing with are large or infinite. Analysis of partial orderings on larger sets therefore requires that we be able to describe specific properties and features of these partial orderings. We have mentioned totally

ordered sets and totally unordered sets. These are what you might call *global properties*
of a partial order—that is, properties that describe the partial order as a whole. Some
elements have a special status with respect to all the other elements in the partially
ordered set.

4.2.11 DEFINITION

Let A be a partially ordered set. Let $x \in A$.

1. The element x is called a **maximal element** for A if there exists no $y \in A$ such
   that $y > x$. Similarly, x is a **minimal element** for A if there does not exist $y \in A$
   such that $y < x$.

2. The element x is the **greatest element** of A if $x \geq y$ for all $y \in A$. Similarly, x is
   the **least element** of A if $x \leq y$ for all y in A.

At first glance, you might think that "greatest element" and "maximal element"
mean the same thing, but, in fact, they are different. One reason why you might confuse
these two is that you are too used to thinking about totally ordered sets!

4.2.12 EXERCISE

Look at the lattice diagrams in Figure 4.21 on page 98. For each determine whether the
partially ordered set has any minimal elements, maximal elements, a greatest element, a
least element. In each case, list all you find. □

4.2.13 EXERCISE

1. Give examples that illustrate the difference between a maximal element and the
   greatest element of a partially ordered set A. Drawing lattice diagrams is a good
   way to do this.

2. Give examples to show that a given partially ordered set can have more than one
   minimal element or none at all. □

4.2.14 THEOREM

Let $\mathcal{T}$ be a totally ordered set and fix $x \in \mathcal{T}$. Then x is a maximal element of $\mathcal{T}$ if and only
if x is the greatest element of $\mathcal{T}$. (That is, in a totally ordered set, the phrases "maximal
element" and "greatest element" mean the same thing!) □

Notice that we have been talking about *the* greatest element of A, as though there
were only one. The following theorem justifies this usage.

4.2.15 THEOREM

Let A be a partially ordered set. Prove that if A has a greatest element, then that greatest
element is unique. (*Hint:* This is a uniqueness theorem. Turn back to page 30 and remind
yourself of the general procedure for proving uniqueness theorems.) □

Remark. Theorems 4.2.14 and 4.2.15 dealt with greatest elements and maximal elements. Analogous statements can be made about least elements and minimal elements. Formulate the appropriate statements, then review the arguments you gave for maximal and greatest elements to see that they can be easily modified to give the analogous results for least and minimal elements.

In general, when we define an object, we cannot assume it is unique just because we wish it so. We are defining the object by means of a property or set of properties. Any objects satisfying those properties will then satisfy the definition. Furthermore, we cannot just add the condition that it must be unique to the definition; if we did, the definition would still not distinguish between the various elements that satisfy the other properties. Thus when we make a definition, we must prove the uniqueness of the object (if indeed it is unique), or accept that there may be more than one object that satisfies the definition.

In addition to global properties, we have *local properties,* that is, properties that refer only to a specific section (subset) of the partially ordered set—they describe only how the partial order behaves "locally" and have nothing to say about the larger picture in the partial order.

4.2.16   DEFINITION

Let A be a partially ordered set. An element x of A is said to be an **immediate successor** of $y \in A$ if $y < x$ and there does not exist an element $z \in A$ such that $y < z < x$.

Likewise, $x \in A$ is said to be an **immediate predecessor** of $y \in A$ if . . . (You should complete this definition for yourself.)

4.2.17   EXERCISE

Show by giving an example that immediate successors and immediate predecessors are not necessarily unique.   □

4.2.18   THEOREM

In a totally ordered set, immediate successors and immediate predecessors (when they exist) are unique.   □

4.2.19   DEFINITION

Let A be a partially ordered set. Let K be a nonempty subset of A. Let $x \in A$.

1. x is an **upper bound** for K if $x \geq y$ for all $y \in K$. If such an element exists, we say that K is **bounded above**.

2. x is called the **least upper bound** of K if
   - x is an upper bound for K, and
   - given any upper bound u for K, $x \leq u$.

In this case, x is denoted by lub K.

3. The least upper bound for K is called the **greatest element** of K if $x \in K$.

4.2.20 EXERCISE

Using Definition 4.2.19 as your model, construct definitions for **lower bound**, **bounded below**, **greatest lower bound**, and **least element** of a subset K of a partially ordered set A.[5] ☐

4.2.21 EXERCISE

Consider $\mathbb{R}$ under the customary ordering $\leq$.

1. Let $K = [-3, 3]$. Find four upper bounds for K. Does K have a least upper bound? A greatest element?

2. Find an example of a subset K of $\mathbb{R}$ in which K has no lower bound.

3. Find an example of a subset K of $\mathbb{R}$ in which K has no least element but has a greatest lower bound. ☐

4.2.22 THEOREM

Let A be a partially ordered set. Let K be a nonempty subset of A. If K has a least upper bound, it is unique. ☐

4.2.23 DEFINITION

A partially ordered set in which every nonempty subset that is bounded above has a least upper bound is said to have the **least upper bound property**.

4.2.24 EXAMPLE

1. You will show in Problem 11 at the end of the chapter that for any set X, $\mathcal{P}(X)$ has the least upper bound property.

2. The most important example of an ordered set with the least upper bound property is $(\mathbb{R}, \leq)$. You should think about this for a while to see if you believe that it is true. It is an axiom of the real number system that $\mathbb{R}$, ordered as usual, has the least upper bound property. (See Chapter 8 for more details.) ■

4.2.25 LEMMA

Let A be a partially ordered set and let K be a subset of A. Define

$$\mathcal{L}_K = \{x \in A : x \text{ is a lower bound for } K\}.$$

---

[5] The least upper bound and greatest lower bound of K are also commonly called the **supremum** and the **infimum** of K. These are denoted by sup K and inf K, respectively.

Suppose that $\mathcal{L}_K$ has a least upper bound. Show that the least upper bound x of $\mathcal{L}_K$ is the greatest lower bound of K.   □

4.2.26   THEOREM

Let A be a partially ordered set that has the least upper bound property. Then every nonempty subset of A that is bounded below has a greatest lower bound. (Or we might say: Every partially ordered set with the least upper bound property also has the **greatest lower bound property**.)

(*Hint:* Use Lemma 4.2.25.)   □

> "Lemma" is another word for theorem, but it has the additional meaning that it is a theorem proved mostly as an aid to proving another, bigger theorem. The lemma may or may not be interesting in its own right.

4.3   Equivalence Relations

Useful relations on a set are often closely related to the way that we organize the set in our minds. For example, the relation "$\leq$" describes the ordering that we place on the real numbers. When we think of sets in everyday life, we may tend to divide them into categories. We divide the set of adult people into two categories: men and women. We divide the set of foods into the four basic food groups: cereals, meats, fruits and vegetables, and dairy products. We divide the set of college students into seniors, juniors, sophomores, and freshmen. We divide the set of integers into even and odd integers.

Such divisions are as useful in mathematics as they are in other endeavors; we will now consider them in mathematical terms. Notice that in each set above, all the categories are mutually exclusive (no member belongs to more than one category) and exhaustive (every member belongs to some category). The mathematical term for this is "partition." A partition of a set S is a collection of mutually exclusive and exhaustive subsets of S. Here are some formal definitions.

4.3.1   DEFINITION

Let S be a set. Let Ω be a collection of subsets of S. The elements of Ω are said to be **pairwise disjoint** if for all elements $A, B \in \Omega$, either $A = B$ or $A \cap B = \emptyset$.

> You may wonder why we do not say "Let A and B be *distinct* elements of K. Then $A \cap B = \emptyset$." This is clearly equivalent and seems less confusing. Our phrasing is chosen to suggest a certain way of thinking about the idea. In practice, when we set out to show that a collection of sets is pairwise disjoint, we usually assume $A \cap B \neq \emptyset$ and then prove that A and B must be the same set.

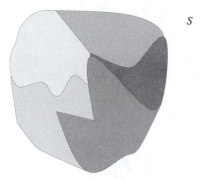

S

Figure 4.9   A partition of a set *S*

4.3.2   EXERCISE

Find an infinite collection of pairwise disjoint subsets of $\mathbb{R}$.   □

4.3.3   DEFINITION

A collection Ω of nonempty subsets of a set S is said to be a **partition of S** provided that the elements of Ω are pairwise disjoint and their union is all of S. That is,

   i. given A and $B \in \Omega$, either $A = B$ or $A \cap B = \emptyset$, and

   ii. $\bigcup_{A \in \Omega} A = S.$

4.3.4   EXERCISE

I said above that a partition of S is a collection of mutually exclusive and exhaustive categories (subsets) of S. Which of the two provisions in the definition corresponds to mutual exclusivity? Which corresponds to the fact that the categories are exhaustive?   □

4.3.5   EXERCISE (Partitions)

Construct the following examples of partitions.

   1. Give an example of a partition of the set $S = \{1, 2, 3, 4\}$.

   2. Give an example of a partition of $\mathbb{N}$ that has four elements.

   3. Give an example of a partition of $\mathbb{R}^2$ that has infinitely many elements.   □

Thus far we have been thinking about dividing a set into disjoint subsets. This division into categories can be seen in another light—in terms of relations. We can think of all members of a given category as being related to each other. This gives a relation; we take all possible ordered pairs of elements that come from the same category.

As you might suspect, the set-theoretic and relational interpretations of categorization are closely related. In fact, *every collection of subsets of A* (not just partitions) can be associated in a natural way with a relation on A. Conversely, *every relation on A* is associated in a natural way with a collection of subsets of A (that is, with a subset of $\mathcal{P}(A)$).

4.3.6 DEFINITION

Let A be a set and Ω any subset of $\mathcal{P}(A)$. If a_1 and a_2 are elements of A, we will say that a_1 is related to a_2

   if there exists an element $R \in \Omega$ that contains both a_1 and a_2.

This relation, $\sim_\Omega$, is called the **relation on A associated with Ω**.

4.3.7 EXERCISE

Let

$$A = \{1, 2, 3, 4, 5, 6\} \quad \text{and}$$
$$\Omega = \{\{1, 3, 4\}, \{2, 4\}, \{3, 4\}\}.$$

List the elements of $\sim_\Omega$.   □

4.3.8 THEOREM

Let A be a set and let Ω be a subset of $\mathcal{P}(A)$. Then the relation $\sim_\Omega$ associated with Ω is symmetric.   □

4.3.9 DEFINITION

Let A be a set. Let $\sim$ be any relation on A. Every $a \in A$ gives us a subset of A:

$$T_a = \{x \in A : a \sim x\}.$$

The set T_a is called the set of **relatives of a under** $\sim$. All of these subsets make up a collection $\Omega_\sim$ of subsets of A:

$$\Omega_\sim = \{T_a : a \in A\}.$$

$\Omega_\sim$ is called the **collection of subsets of A associated with** $\sim$.

4.3.10 EXERCISE

Let $A = \{1, 2, 3, 4, 5, 6\}$. Suppose

$$\sim = \{(1, 1), (2, 2), (2, 3), (2, 5), (3, 5), (4, 2), (4, 3), (4, 5), (5, 2), (5, 3), (5, 5)\}.$$

   1. For each $a \in A$, find T_a.
   2. Now find $\Omega_\sim$.   □

4.3.11 PROBLEM

Let $A = \{1, 2, 3, 4, 5, 6\}$.

   1. Consider the following subset of $\mathcal{P}(A)$:

$$\Omega = \{\{1, 2, 3, 4\}, \{5, 6\}\}.$$

   Find $\sim_\Omega$.

2. Consider the following relation on A:

$$\sim = \{(1, 1), (2, 2), (3, 3), (4, 4), (5, 5), (6, 6), (1, 2),$$
$$(1, 4), (2, 1), (2, 4), (4, 1), (4, 2), (3, 6), (6, 3)\}.$$

Find $\Omega_\sim$.   □

4.3.12   PROBLEM

Paul and Bettie have four sons: P. W. (age 12), Pat (age 10), Will (age 7), and Ben (age 4). In this family, person A is related to person B if A is older than B. (Bettie is older than Paul.)

What is the collection of subsets associated with the relation?   □

4.3.13   PROBLEM

Poll between 5 and 10 people. From this set of people, form the following subsets.

• The set of all people who own cats.

• The set of all people who own gerbils.

• The set of people who do not have pets.

Describe the relation associated with this collection of subsets. Theorem 4.3.8 says that it should be symmetric. Is it reflexive? Antisymmetric? Transitive? Explain.   □

4.3.14   PROBLEM

Give an example of a set A and a subset Ω of $\mathcal{P}(A)$ such that

1. the relation $\sim$ associated with Ω is reflexive.
2. the relation $\sim$ associated with Ω is not reflexive.   □

Complete the statement of the following theorem, then prove it.

4.3.15   THEOREM

Let A be a set and let Ω be a subset of $\mathcal{P}(A)$.

If _____, then the relation $\sim_\Omega$ is reflexive.   □

4.3.16   THEOREM

Let A be a set and let Ω be a subset of $\mathcal{P}(A)$. Suppose that the elements of Ω are pairwise disjoint. Then the relation $\sim_\Omega$ associated with Ω is transitive.   □

We can now gather several smaller results into one bigger one:

4.3.17   COROLLARY

Let S be a set. Let Ω be a partition of S. Then the relation on S associated with Ω is reflexive, symmetric, and transitive.   □

"Corollary" is another word for theorem, but it has the additional meaning that its proof follows immediately from previously proved theorems.

4.3.18   DEFINITION

A relation on S that is reflexive, symmetric, and transitive is called an **equivalence relation**.

4.3.19   EXERCISE (Equivalence relations)

Give an example of an equivalence relation. (Be sure to specify the set on which the relation is defined.)   □

4.3.20   LEMMA

Suppose A is a set. Let $\sim$ be an equivalence relation on A and let $a, b \in A$. Then

$$T_a = T_b \text{ if and only if } a \sim b.$$ □

We have seen that a partition of S yields an equivalence relation. Conversely, we have the following.

4.3.21   THEOREM

Let $\sim$ be an equivalence relation on a set S. Then $\Omega_\sim$ forms a partition of S. That is,

- $\bigcup_{x \in S} T_x = S$, and
- for x and y in S, either $T_x = T_y$, or $T_x \cap T_y = \emptyset$.

(*Hint:* This is a set-theoretic theorem. You are showing that sets are equal, so you will need to use element arguments just as you did in the chapter on set theory. For additional guidance on the proof of the second part, see the box on page 78.)   □

By now it should be pretty clear that the study of equivalence relations on S is intimately related to the partitioning of S into disjoint subsets. In fact, *they are two faces of the same problem.* When we study partitions, we can view them mathematically in either way: We can bring the tools of set theory to bear when we think of a partition as a collection of subsets, or we can use what we know about relations to make sense of them! These two views complement each other, and it is useful to be able to cross over easily from one to the other.

To make this juggling act advance more smoothly, we introduce some more language.

4.3.22 DEFINITION

Let S be a set, let $\sim$ be an equivalence relation on S, and let $x \in S$. Then T_x is called the **equivalence class of x under $\sim$.**

Following this, $\Omega_\sim$ is called the set of equivalence classes of S given by $\sim$ (or simply the **equivalence classes of $\sim$.)**

> Lemma 4.3.20 and Theorem 4.3.21 explain why we introduced this new set of terms. Theorem 4.3.21 tells us that the T_x's are a set of mutually exclusive and exhaustive "categories" of S. Lemma 4.3.20 tells us what these "categories" (equivalence classes) are. Two elements of S fall into the same "category" (equivalence class) if and only if they are related to each other. All the elements in an equivalence class are related to one another. Any element outside a particular equivalence class is unrelated to elements in the equivalence class.

4.3.23 EXERCISE

Show that the following relations $\sim$ on the specified set S are equivalence relations. In each case do this in two ways:

- By identifying the equivalence classes and noting that they partition S.
- By showing directly that $\sim$ is reflexive, symmetric, and transitive.

1. $S = \{p : p$ is a person in Ohio$\}$. $A \sim B$ if A and B were born in the same year.
2. $S = \mathbb{Z}$. $a \sim b$ if $|a| = |b|$.
3. $S = \mathbb{Z}$. $a \sim b$ if $a - b$ is an integral multiple of 5.   □

4.4 Graphs

Consider the digraph shown in Figure 4.3. Notice that it is possible to get from Dubuque to Chicago and back. However, it is only possible to go one way between Rochester and Philadelphia. If every possible route was two-way, it would make sense to eliminate the two arcs going different ways and replace them by a single undirected edge. The resulting diagram would become Figure 4.10.

If we think about what happened in mathematical terms, we started with a symmetric relation in which no element was related to itself. In such a relation, for distinct x and y, saying that x is related to y is the same as saying that y is related to x; therefore, it is sometimes convenient to simply conjoin the two ordered pairs (x, y) and (y, x) and think of them as a single *un*ordered pair $\{x, y\}$ (in effect, the two-element set containing x and y). So a symmetric relation in which no element is related to itself can be thought of as a set of unordered pairs.

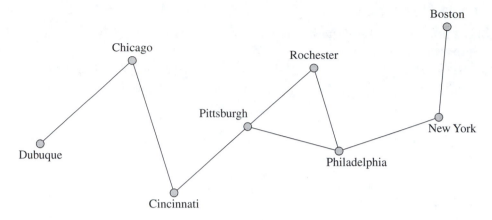

Figure 4.10   Two-way transportation network

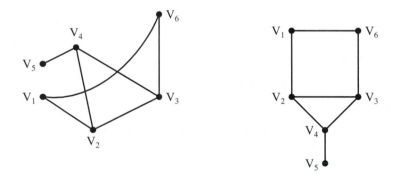

Figure 4.11   A graph

4.4.1   DEFINITION

Let V be a set and E a set of unordered pairs of elements of V. Then the pair $G = (V, E)$ is called a **graph** on the set V.

The elements of the set V are called the **vertices of G**.

If v_1 and v_2 are vertices of G, and $\{v_1, v_2\} \in E$, we say that there is an **edge** in G joining v_1 and v_2. Thus the elements of the set E are called the **edges of G**.

Figure 4.11 shows a convenient way to represent graphs. The diagram shows the vertices and the edges between them. The placement of the vertices and the lengths of the edges do not matter. Crossings where there is not a vertex have no significance and it makes no difference whether the edges are drawn as straight lines or curves. Indeed, the left and the right diagrams represent the same graph.

4.4.2 EXERCISE

For the graph shown in Figure 4.11, list the vertices and the edges. Verify that both diagrams give the same answer. □

Graphs have many applications. Here are some examples that illustrate the variety of fields in which graph theory can be useful.

4.4.3 EXAMPLE

1. The two graphs shown in Figure 4.12 represent the chemical bonds of propane and isobutane molecules. (The vertices that are attached to four edges are carbon atoms, the vertices that are attached only to one edge are hydrogen atoms.)

2. Figure 4.13 shows the family tree of Indo-European language groups, as proposed by August Schleicher in 1862. English is part of the Germanic group. (Modern versions of the diagram are more complex and include additional groups unknown to Schleicher.)

3. Figure 4.14 shows an electrical resistor network. The vertices in the network are connected by imperfect conductors, called resistors. (Similar graphs can be drawn showing other kinds of electrical networks and computer networks.) ■

Remarks. Graph theory is a rich and varied subject. This will be a very brief treatment. To focus the discussion, I have decided to introduce only those ideas and theorems that I need to make it possible to discuss map coloring. In particular, I will limit the sorts of graphs that we look at.

1. Though the definition of graph makes perfect sense if the set of vertices is infinite, we will confine our discussion to talking about finite graphs. So whenever I say "Let $G = (V, E)$ be a graph" I really mean "Let V be a finite set and let $G = (V, E)$ be a graph on V."

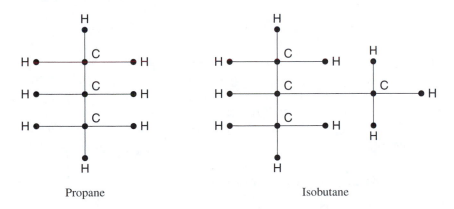

Propane   Isobutane

Figure 4.12   Bond graphs

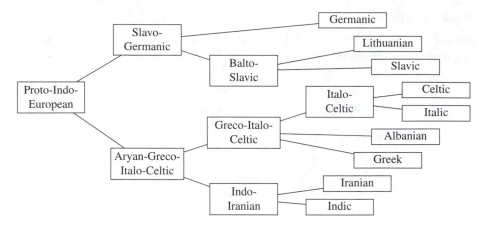

Figure 4.13   The descent of Indo-European languages

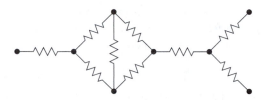

Figure 4.14   An electrical resistor network

2. Some discussions of graphs assume the possibility of an edge from a vertex to itself and of more than one edge joining the same pair of vertices. The definition I gave does not allow for either of these.

It would be an understatement to call graph theory a "definition-rich" subject. That is, a there are lots of words you need to know just to carry on a conversation about graph theory. Please bear with me while I define lots of terms. The good news is that most of the ideas are intuitive, and the definitions are easy to remember. (As you read, draw lots of pictures to help you understand the meanings of the terms.)

> **Something to keep in mind:** Because the graphs we will be dealing with have a finite number of vertices and, therefore, a finite number of edges, mathematical induction (on either the number of vertices or the number of edges) is very well suited to proving theorems about these graphs.

4.4.4   DEFINITION (Inside the graph)

Let $G = (V, E)$ be a graph.

1. Let u and v be vertices in G. If the pair $\{u, v\}$ is in E, then u and v are said to be **adjacent vertices**. The edge $\{u, v\}$ is said to be **incident** with the vertices u and v.

2. The **degree** of a vertex v is the number of edges incident with v. The degree of v is denoted by $\deg(v)$.

3. A vertex of degree one is called an **end vertex.**

4. If u, v, and w are vertices, and $\{u, v\}$, $\{v, w\}$, and $\{w, u\}$ are all edges, then u, v, and w are said to form a **triangle** in G.

5. Let $V^* \subseteq V$. If $E^* \subseteq E$, and the edges in E^* are incident only with vertices in V^*, then $G^* = (V^*, E^*)$ is said to be a **subgraph** of G.

6. Let $V^* \subseteq V$, and let E^* be the set of all edges that join pairs of vertices in V^*. Then $G^* = (V^*, E^*)$ is called the **subgraph of** G **generated by** V^*.

4.4.5 EXERCISE

For this exercise, refer once again to Figure 4.11. Answer the following questions about the graph G depicted there.

1. Find a pair of vertices of G that are adjacent and a pair of vertices that are not adjacent.

2. What edges of G are incident with v_3?

3. Which vertex in G has largest degree? Which vertex has smallest degree?

4. Does the G contain a triangle?

5. Draw the subgraph of G that is generated by the vertices $\{v_1, v_2, v_4, v_5\}$.

6. Draw the subgraph of G that is generated by the vertices $\{v_1, v_3, v_5, v_6\}$.

7. Find a subgraph of G that is *not* the subgraph generated by its set of vertices. $\square$

4.4.6 THEOREM

Let $G = (V, E)$ be a graph. Let e be the number of edges of G. Then

$$\sum_{v \in V} \deg(v) = 2e. \qquad \square$$

4.4.7 DEFINITION (Moving around the graph)

Let $G = (V, E)$ be a graph.

1. Let u and v be vertices in G. A **walk** in G from u to v is an alternating list of vertices and edges in which each edge is incident with the vertices that come before and after it:

$$u, \{u, v_1\}, v_1, \{v_1, v_2\}, v_2, \{v_2, v_3\}, v_3, \ldots, v_{n-1}, \{v_{n-1}, v\}, v.$$

We will denote such a walk by

$$u \to v_1 \to v_2 \to \cdots \to v_{n-1} \to v.$$

The **length** of the walk is the number of edges in the walk. (In this case, the walk has length n.)

2. A **path** from a vertex u to a vertex v is a walk in which no vertex appears more than once.

3. A **cycle** in a graph is a walk

      (a) that starts and ends at the same vertex v.

      (b) in which no vertex other than v appears more than once.

      (c) in which no edge is used more than once.

(A cycle is sometimes called a *closed path*.)

> **Induction Tip:** Suppose you are doing a proof by induction on the number of edges in a graph. You first show that the theorem is true for a graph with one edge and then assume it is true for all graphs with k (or fewer) edges. During the induction step, you assume you have a graph with $k + 1$ edges and ask yourself what happens if you remove an edge. In effect, you are hoping to apply the induction hypothesis to the subgraph with the same vertices as the original graph and all edges except the removed edge.
>
>   Likewise, if you are doing induction on the number of vertices, your induction step starts with a graph $G = (V, E)$, where $V = \{v_1, v_2, \ldots, v_k, v_{k+1}\}$. You are interested in the subgraph *generated* by the vertices $\{v_1, v_2, \ldots, v_k\}$.

4.4.8 EXERCISE

Provide examples, in the form of well-labeled diagrams, of the following mathematical objects:

1. A graph and a walk in the graph that is not a path.

2. A graph and a walk in the graph that starts and ends at the same vertex v and in which no vertex other than v appears more than once, but that isn't a cycle.

3. A graph with no cycles.

4. A graph whose shortest cycle has length 4 and whose longest cycle has length 7.

          □

4.4.9 THEOREM

Let $G = (V, E)$ be a graph. Let u and v be vertices of G. Then there is a walk in G from u to v if and only if there is a path in G from u to v.

    (*Hint:* Proceed by induction on the length of the walk.)   □

4.4.10 DEFINITION (Connectedness in graphs)

Let $G = (V, E)$ be a graph.

1. A graph $G = (V, E)$ is said to be **connected** if for every pair of vertices u and v there is a path in G from u to v.

   A graph that is not connected is said to be disconnected.

2. A **tree** is a connected graph with no cycles.

4.4.11 EXERCISE

1. Give an example of a connected graph and of a disconnected graph. Use your own examples to explain why it makes sense to say that a graph is connected if there is a path joining any two vertices and disconnected otherwise.

2. Give three different examples of trees with seven vertices.   □

4.4.12 THEOREM

Let $G = (V, E)$ be a graph. Then G is connected if and only if there is a walk that goes through all the vertices.   □

4.4.13 EXERCISE

Show by giving an example that there are connected graphs in which there is no path that goes through all the vertices.   □

4.4.14 LEMMA

Let $G = (V, E)$ be a connected graph. Let $e \in E$. Then $G' = (V, E \setminus \{e\})$ is a connected graph if and only if G has a cycle containing e.   □

Suppose that $G = (V, E)$ is a graph and that $e \in E$. If G has no cycle containing e, then Lemma 4.4.14 tells us that removing e from G will disconnect G. Problem 19 at the end of the chapter tells us that removing e from G breaks G into exactly two connected pieces. (The proof is a bit long, but not particularly tricky.) First make sure you understand and believe this, then feel free to use the fact in the proof of the next theorem.

4.4.15 THEOREM

Let $G = (V, E)$ be a graph with n vertices and e edges. Show that the following statements are equivalent.

1. G is a tree.
2. G is connected and $n = e + 1$.
3. G has no cycles and $n = e + 1$.
4. If u and v are vertices in G, then there exists a unique path connecting u and v.

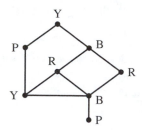

Figure 4.15   A 4-coloring of G

5. G has no cycles, but the addition of any edge to G will create a graph with a cycle. (By adding an edge to G, I mean joining by an edge a pair of nonadjacent vertices of G.) □

4.4.16 THEOREM

Let $T = (V, E)$ be a tree with n vertices. Then T has at least two end vertices. □

4.4.17 DEFINITION (Graph coloring)

Let $G = (V, E)$ be a graph. Suppose we assign a color to each vertex of G in such a way that adjacent vertices are colored with different colors. If such an assignment can be made using at most k colors, then we call it a k-**coloring** of G and we say that G is k-**colorable**.

The smallest number k such that G is k-colorable is called the **chromatic number of G** and is denoted by $\chi(G)$.

4.4.18 EXAMPLE

Figure 4.15 shows a 4-coloring of a graph, G. The four colors, red, yellow, blue, and purple, are depicted by the letters R, Y, B, and P, respectively. Can you find a 3-coloring? A 2-coloring? What is $\chi(G)$? ■

4.4.19 EXERCISE

1. Give an example of a graph that is 5-colorable but not 4-colorable.

2. Give an example of a graph with no triangles that is not 2-colorable. □

4.4.20 EXERCISE (Scheduling problems)

Graph coloring has many applications. For instance, this simple exercise will show you how it can be used to solve scheduling problems.

The Smooth-Talking Advertising Agency has people working on five ad campaigns. They need to make presentations to all five clients on Monday and Tuesday mornings. A presentation takes an entire morning and everyone on the ad team is needed for the presentation. Suppose they have their executives working on the following teams.

Account	Team			
Aquatic Incidents, Inc.	Carlton	Shouse	Benjamin	Dow
Billy Bob's Buffalo Bar-B-Que	Wade	Wallace	Patrick	Smith
Coalition for Lost Causes	Wayne	Carlton	Smith	Failey
Devil's Island Tourist Authority	Wade	Dow	Wallace	Patrick
Extreme Measures, Ltd.	Failey	Wayne	Cunningham	Dill

1. Draw a graph in which each vertex is a client and draw edges between clients if the same executive is working on both campaigns.

2. Use graph coloring to help you schedule the presentations of the ad campaigns. □

4.4.21 THEOREM

Let $G = (V, E)$ be a graph in which every $v \in V$ has degree less than or equal to Δ. Then G is $(\Delta + 1)$-colorable. □

4.4.22 EXERCISE

The bound given in Theorem 4.4.21 can be achieved, but need not be.

1. Give an example of a graph G with a maximum vertex degree of Δ for which $\chi(G)$ is equal to $\Delta + 1$.

2. Give an example of a graph G with a maximum vertex degree of Δ for which $\chi(G)$ is smaller than $\Delta + 1$. □

4.4.23 DEFINITION (Planarity)

Let $G = (V, E)$ be a graph. Then G is said to be **planar**, if G can be drawn in the plane in such a way that the edges do not cross. Such a drawing of G is called a **plane drawing**.

Remark. Note that the definition of planar does not say that *none* of the many possible drawings of G may have crossings. Only that there is at least one without crossings. Figure 4.11 shows a planar and a nonplanar drawing of the same graph. The graph is planar because it is possible to draw it in the plane without any crossings.

The question of whether a particular graph is planar can be very tricky. (If you can actually exhibit a planar drawing of the graph, of course, this settles the issue. But what if you cannot? This may simply mean that you have not been clever enough to find one.) I will skirt the issue of how to determine whether a graph is planar by simply giving a few examples and moving on. You should "play" with the examples enough to convince yourself that what I say is true.

4.4.24 EXAMPLE

1. All trees are planar.

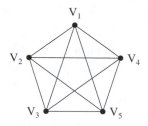

Figure 4.16 K_5

2. The graph shown in Figure 4.16 is called K_5; it is nonplanar. K_5 is the **complete graph** on 5 elements, since there are five vertices and every pair of vertices is joined by an edge. Likewise, we can talk about the complete graph on 3, 4, or 17 elements. (What would these look like? Which are planar?)

3. Any graph that contains a nonplanar graph as a subgraph fails to be planar.

4. Consider a standard map (real or imagined) of a continent divided into countries. If we put a vertex at the capital of each country and draw an edge between capitals of countries that share a boundary, the resulting graph is planar. If the edges are drawn so that they stay inside the two countries in question except when the edge crosses the shared boundary, the resulting drawing is a plane drawing. (See Figure 4.17.)   ∎

4.4.25   EXERCISE

Verify that the graph shown in Figure 4.17 (page 93) is the graph derived from the map in the same figure. (Identify each of the cities that appear on the map as one of the vertices on the graph. Then check the edges.) Find a 4-coloring of the graph.   □

4.4.26   EXERCISE

Consider the map of South America shown in Figure 4.18 (page 94). Exhibit a planar drawing of the graph that joins the capitals of countries that share a boundary. Find a 4-coloring of this graph.   □

4.4.27   EXAMPLE (Faces of a plane drawing)

Figure 4.19 (page 95) shows two plane drawings of the same graph. Each breaks up the plane into five disjoint regions called *faces*. In each case, there are several bounded regions and one unbounded region. It is possible to draw the graph so that any one of the faces is unbounded. Note that in the first drawing face 5 is unbounded, whereas in the second drawing face 2 is.

Despite the fact that the idea of a face is a fairly intuitive one, actually **defining** a face rigorously is not so easy. We are not going to do it. In order to maintain the rigor of our discourse, I will set out certain facts about faces that we will assume for the coming discussion. Like faces themselves, these facts are fairly intuitive, and you should have

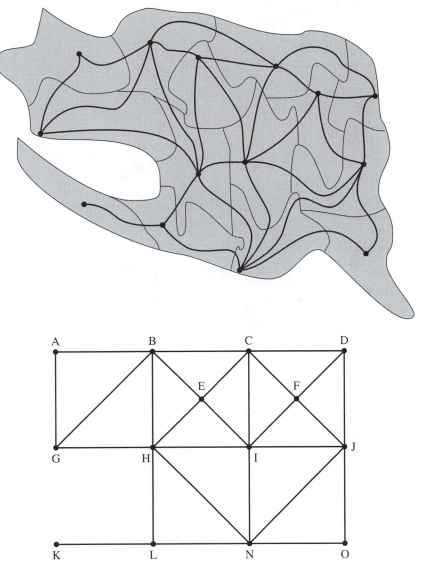

Figure 4.17   A graph of a map

no trouble convincing yourself of their plausibility. Keep in mind, however, that there is a break in the chain of logic here and that to really set things straight requires a careful definition of face followed by proofs of the assumed facts.   ■

Assumptions.   Let $G = (V, E)$ be a planar graph with v vertices and e edges. Suppose we have a plane drawing of G that has f faces.

Figure 4.18   South America

1. If G is connected and has one edge, then $f = 1$.
2. If G is a tree, then $f = 1$.
3. Removing an edge from a cycle in G decreases the number of faces by 1.
4. If G is connected and $v \geq 3$, then $3f \leq 2e$. (It takes at least three edges to bound a face. Each edge bounds either one or two faces.)

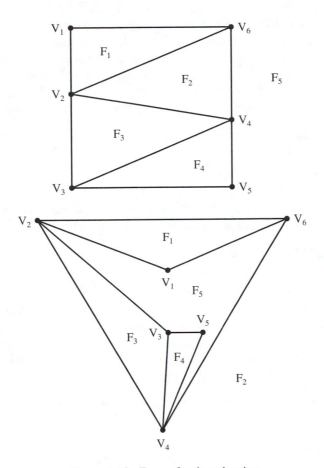

Figure 4.19 Faces of a plane drawing

4.4.28 THEOREM (Euler's formula)

Let G be a connected planar graph. Fix a plane drawing of G. Let v, e, and f be the number of vertices, edges, and faces, respectively, of G in this plane drawing. Then

$$v + f - e = 2. \qquad\qquad \square$$

4.4.29 THEOREM

Every planar graph has a vertex of degree at most five. (*Hint:* Use Euler's formula.) $\square$

Coloring Maps

Cartographers have long known that only four colors are needed to paint a map in such a way that countries that share a boundary are colored in different colors. A choice of colors that meets this criterion is called a "coloring" of the map. The graph colorings you found in Exercises 4.4.25 and 4.4.26 give such colorings for the map of South America and for the imaginary map shown in Figure 4.17—just paint each country with the color you chose for the vertex representing it. Likewise, if we have a map coloring, we get a coloring of the corresponding graph. Finding a way to color a map and finding a graph coloring of the planar graph derived from the map are, in fact, the same problem. It is fairly straightforward to prove that coloring any arbitrary map can be accomplished with no more than six colors.

4.4.30   THEOREM

Every planar graph is 6-colorable.   □

An elaboration of this argument will prove that map coloring requires no more than five colors. (But the argument requires some cleverness; you will have to work at it.)

4.4.31   THEOREM

Every planar graph is 5-colorable.   □

So where does that leave the cartographer's empirical knowledge that we should be able to color every planar graph with no more than four colors? The "Four Color Map Theorem" is very famous because the mathematical question remained unresolved for more than a century. It was finally solved in the mid-1970s using a computer. See the Questions to Ponder for some additional information.

■ PROBLEMS

1. Suppose you have any set A with n elements and any set B with m elements. Can you guess how many elements $A \times B$ has? (Start by working out some examples. Make a conjecture and check your answer against one or two more examples. Refine your guess, if necessary.) Prove your conjecture, proceeding by induction on m.

2. What might a graphical representation of the relation $\leq$ on $\mathbb{R}$ look like? (This should conjure up a familiar picture!)

3. Consider the relation on $\{a, b, c, d\}$ that is represented graphically in Figure 4.20.

   (a) Is the relation reflexive? Symmetric? Antisymmetric? Transitive? (Can you describe each property in graphical terms?)

   (b) Write out the set of pairs represented by the relation.

   (c) Let $A = \{a, b, c\}$. $\sim = \{(a, a), (b, b), (c, c), (a, b), (b, c), (a, c)\}$. Draw a graph of this relation. Is the relation reflexive? Symmetric? Antisymmetric? Transitive?

   (d) Repeat the process with $A = \{a, b, c, d, e\}$.

$$\sim = \{(a, a), (b, b), (c, c), (d, d), (e, e), (a, b), (b, a),$$
$$(b, c), (c, b), (c, a), (a, c), (d, e), (e, d)\}.$$

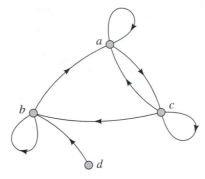

Figure 4.20

4. Mathematical economists consider entities called *consumers*. (A consumer may be an individual, a household, or even a larger group that has some common economic purpose.) The role of a consumer is to decide on a consumption plan—a decision about what commodities to acquire, what commodities to produce, and how much of each. When "Charlie, the consumer" considers two possible consumption plans, P_1 and P_2, he can come to one of three conclusions:

   i. Charlie prefers P_1 to P_2.

   ii. Charlie prefers P_2 to P_1.

   iii. Charlie is indifferent to whether he chooses P_1 or P_2.

For Charlie, we can get three relations on the set of possible consumption plans: The relation "is preferred to," the relation "is at least as desired as," and the relation "is indifferent to."

    Economists like to assume that their consumers are rational beings, so they make certain assumptions about these relations.

- One of the three relations they assume to be reflexive, symmetric, and transitive.
- One relation they assume to be reflexive and transitive.
- One relation they assume to be only transitive.

Which assumptions do you think should correspond to each relation, and why? Why don't economists assume that any of these relations are antisymmetric?

5. Are there any sets A for which $(\mathcal{P}(A), \subseteq)$ *is totally ordered? Prove your answer.*

6. In Exercise 4.2.7, you showed that partially ordered sets can have totally ordered subsets. It is perhaps more surprising that any partial ordering on a finite set is contained in a total ordering on the same set.[6] Let us try to get a sense for why this is true by looking at a couple of examples. Construct total orderings that contain the partial orderings shown in Figure 4.21. (*Hint:* Remember that orderings are collections of ordered pairs. "Contained in" means "is a subset of." That is, we are adding new relationships—pairs—without getting rid of old ones. You will have to decide on an order for elements that were previously incomparable without disrupting the previous order relationships. There are many correct ways to do this.)

---

[6] The proof of this is challenging, but I encourage you to give it a try. You have all the mathematical tools that you need to carry it through. I provide some hints in "Questions to Ponder."

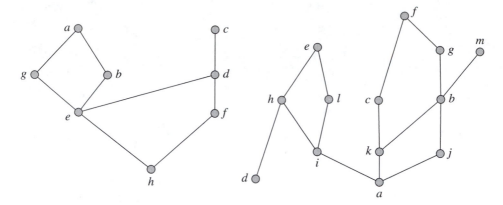

Figure 4.21   Find total orderings that contain these partial orderings.

7. Is it possible for a partially ordered set A to have both a least element and a minimal element that is *not* a least element? Prove your answer.

8. This problem asks you to construct some examples. It might be useful for you to know that for the first two examples you have familiar examples at your fingertips. For the last two you may have to be a bit creative.

(**a**) Give an example of a totally ordered set in which every element has an immediate successor.

(**b**) Give an example of a totally ordered set in which no element has an immediate successor.

(**c**) Give an example of an (infinite) totally ordered set in which all but the least element and the greatest element have an immediate predecessor (and they do not).

(**d**) Construct an example of a partially ordered set in which every element has an immediate successor but there are infinitely many elements that have no immediate predecessor.

9. Look at the right-hand lattice diagram shown in Figure 4.21. Consider the following subsets:

i. $\{h, l\}$ **ii.** $\{c, h, l\}$ **iii.** $\{e, h, i, l\}$ **iv.** $\{k, j\}$

   For each subset:

(**a**) Find the set of upper bounds (it may be empty).

(**b**) If the set of upper bounds is nonempty, determine whether there is a least upper bound and, if so, whether it is a greatest element.

(**c**) Do the same for lower bounds, greatest lower bounds, and least elements.

10. Let m and n be natural numbers. We say that **n divides m** provided that there exists $a \in \mathbb{N}$ such that $m = an$. (We denote this relation by $n \mid m$.)

(**a**) Prove that $\mathbb{N}$ is partially ordered under the relation $\mid$.

(**b**) Is $\mid$ a *total* order on $\mathbb{N}$? Explain.

(**c**) Draw a lattice diagram that depicts the order $\mid$ on the set $\{1, 2, 3, \ldots, 15\}$.

(**d**) Does $\{2, 3, 4, 5, \ldots\}$ have any minimal or maximal elements (with respect to the order $\mid$)?

(e) Complete the following sentence:

> An element $d \in \mathbb{N}$ is a lower bound (respectively, upper bound) for $S \subseteq \mathbb{N}$ under the order $|$ if and only if _____.

(f) Show that the set $\{12, 18\}$ has a greatest lower bound and a least upper bound in $(\mathbb{N}, |)$. What are they? (There are more common names for these. Do you know what they are?)

11. Consider $\mathcal{P}(\mathbb{N})$ under the partial ordering $\subseteq$.

   (a) Give an example of a nonempty subset of $\mathcal{P}(\mathbb{N})$ with no greatest element.

   (b) Let $K = \{\{2, 3, 4, 12\}, \{3, 6, 9, 12\}, \{1, 2, 3, 4, 5, 6, 7, 8, 9, 10, 11, 12\}\}$. Find three upper bounds for K and three lower bounds for K in $\mathcal{P}(\mathbb{N})$. Does K have a least upper bound? Does K have a greatest lower bound?

   (c) Let X be any set. Suppose that K is a nonempty subset of $\mathcal{P}(X)$—ordered as usual under $\subseteq$. How would you construct the least upper bound of K? How about the greatest lower bound of K?

12. Let A be a partially ordered set. Suppose $X \subseteq Y \subseteq A$.

   (a) Assuming that all the least upper bounds and greatest lower bounds exist, prove that

   $$\text{glb}(Y) \leq \text{glb}(X) \leq \text{lub}(X) \leq \text{lub}(Y).$$

   (b) Find two subsets X and Y of $\mathbb{R}$ for which X is a *proper* subset of Y and yet

   $$\text{glb}(Y) = \text{glb}(X) \text{ and } \text{lub}(X) = \text{lub}(Y).$$

13. The least upper bound property requires that every *nonempty* subset that is bounded above have a least upper bound. The qualification that the set be nonempty is important. To see why, consider $\emptyset$ as a subset of $\mathbb{R}$. Describe the set of upper bounds of $\emptyset$. (Take care, the set of upper bounds is *not* empty!) Is there a least upper bound?

14. Consider the relation "$\leq$" on $\mathbb{R}$. Let $r \in \mathbb{R}$. Describe the set of relatives of r. What is $\Omega_{\leq}$? (*Hint:* Use interval notation when you state your answer.)

15. Peter is throwing a dinner party to introduce his sister—Sarah—to his friends—Karen, Donna, Keith, and Josh. Peter, as the host, knows everyone's name and everyone knows his. Sarah knows no one except Peter. Karen knows Keith's and Josh's names. She met Donna once, but has forgotten her name. Donna remembers Karen and she also knows Josh. Keith knows Karen and Josh. Josh knows Donna and Keith.

   What is the collection of subsets associated with the relation "knows the name of" on the set of people at the dinner party?

16. Show that these relations are equivalence relations. Describe the equivalence classes.

   (a) $S = \mathbb{R}$. $a \sim b$ if $a - b$ is an integer.

   (b) $S = \mathbb{R}$. $a \sim b$ if $\lfloor a \rfloor = \lfloor b \rfloor$ (where $\lfloor x \rfloor$ is the greatest integer less than or equal to x. For example:

   $$\lfloor 2.3 \rfloor = 2$$
   $$\lfloor -1.5 \rfloor = -2$$
   $$\lfloor 3 \rfloor = 3.)$$

   (c) $S = \mathbb{R}^2$. $(x_1, y_1) \sim (x_2, y_2)$ if $x_1^2 + y_1^2 = x_2^2 + y_2^2$.

(d) $S = \mathbb{R}^2$. $(x_1, y_1) \sim (x_2, y_2)$ if $3x_1 + 2y_1 = 3x_2 + 2y_2$. (*Notice:* There is nothing "special" about the numbers 3 and 2. If we pick arbitrary parameters a and b, what will the equivalence classes look like?)

17. Let Ψ be the following subset of $\mathbb{Z} \times \mathbb{Z}$:

$$\Psi = \{(a, b) : a, b \in \mathbb{Z}, b \neq 0\},$$

and define $\sim$ on Ψ by $(a, b) \sim (c, d)$ if and only if $ad = bc$. Prove that $\sim$ is an equivalence relation. Describe the equivalence class of $(1, 2)$. Describe the equivalence class of $(3, 4)$. Describe the equivalence class of (a, b). (This is one of the most important and fundamental equivalence relations in mathematics. Do you recognize it?)

18. Let $G = (V, E)$ be a graph on $2k$ vertices containing no triangles. Show, by induction on k, that G has at most k^2 edges. Give an example of a graph for which this upper bound is achieved.

19. Let $G = (V, E)$ be a connected graph, let $e = \{u, w\} \in E$, and let $G^* = (V, E \setminus \{e\})$. Suppose that there is no cycle in G containing e. Let

$$V_1 = \{v \in V : \text{ there is a path in } G^* \text{ from } u \text{ to } v\},$$

and let $V_2 = V \setminus V_1$. Finally, let $H_1 = (V_1, E_1)$ be the subgraph of G^* generated by V_1. Likewise, let $H_2 = (V_2, E_2)$ be the subgraph of G^* generated by V_2.
   Do the following:

   (a) Prove that if $k \in V$ is joined to a vertex in V_1 by a path in G^*, then $k \in V_1$.

   (b) Show that H_1 and H_2 are connected graphs.

   (c) Show that $E_1 \cap E_2 = \emptyset$, and $E \setminus \{e\} = E_1 \cup E_2$.

   Then we can conclude that $G^* = (V_1 \cup V_2, E_1 \cup E_2)$. We say that G^* is the *disjoint union of H_1 and H_2* since the set of vertices of G^* is the disjoint union of the sets of vertices of H_1 and H_2, and likewise for the sets of edges. H_1 and H_2 are called the **connected components of G.**

20. Any disconnected graph is the disjoint union of finitely many connected subgraphs called **connected components**. Using ideas set out in Problem 19 as your inspiration, define the notion of connected component for graphs. Suppose that $G = (V, E)$ is a graph and $v \in V$. Give a mathematical description for the connected component of G that contains v.

21. Let $G = (V, E)$ be a graph.

   (a) Let $v \in V$ be an end vertex. Then the subgraph generated by $V \setminus \{v\}$ is connected.

   (b) Suppose G is a connected graph with n vertices. Then G has at least $n - 1$ edges. (*Hint:* Prove first the case when every vertex in G has degree at least two.)

■ QUESTIONS TO PONDER

1. In Problem 6, I said that every partial ordering on a finite set is contained in a total ordering. Can you prove this? (*Hint:* You will proceed by induction on the size of the underlying set. Look back at Problem 6. Try to systematize your approach so that if you take away a point and then construct a total order on the remaining points, you will easily be able to describe how

the point you removed should be related to the ones that remained. It will simplify matters if you choose with some care what sort of point to remove in the first place.)

2. You may ask yourself:

   ▪ What happens when we start with a collection of sets, get a relation, and then create a collection of sets from the relation?

   ▪ What happens when we start with a relation, get a collection of sets, and, in turn, create a relation from the resulting collection of sets?

   This is a fairly interesting question to explore and (better yet) a tractable one.

   Since every collection of sets yields a relation and every relation a collection of sets, you should be able to get a collection of subsets of A from the results of Exercise 4.3.7 and the first part of Problem 4.3.11. Likewise, you should be able to construct a collection of pairs from your answer to Exercise 4.3.10 and the second part of Problem 4.3.11. *Do that now.*

   What do you observe in this game of "back and forth"? Speculate about what is going on. Make conjectures. Prove theorems. (The notation necessary to write this stuff down can be a bit daunting. Hang in there!)

3. Consider the situation described in Problem 4.

   (a) As a way of helping him make a decision, it may seem reasonable to ask whether the set of consumption plans from which Charlie must choose is at least partially ordered. Is this so? Explain.

   (b) You should have figured out that the relation "is indifferent to" is an equivalence relation on the set of consumption plans for Charlie. Why might it be reasonable to call the equivalence classes given by this relation "indifference classes"?

   (c) Though the set of consumption plans is not partially ordered, the set of indifference classes is. Think about what this would mean and why it makes some sense.

   (d) How might a knowledge of the partial ordering on the set of "indifference classes" help Charlie to make a decision?

4. Prove that a graph is 2-colorable if and only if it does not have a cycle of odd length. (This theorem is called König's theorem.)

5. Show that for any integer k, there is a graph G that contains no triangles and such that $\chi(G) = k$.

6. We have talked about the "planarity" of a graph. What about trying to draw graphs on other surfaces without crossings? For instance, show that K_5, the complete graph on five vertices, can be drawn on the surface of a torus (a doughnut) without crossings. Can you find other graphs that are not planar but can be drawn without crossings on the surface of a torus? Can you find an analog to Euler's formula for graphs that can be drawn without crossings on the surface of a torus? What about other surfaces? What can you say about a sphere? What about a two-handled torus (a doughnut with two holes)?

7. (a) Try to write a careful mathematical definition for the "face" of a planar graph. If you can't quite manage it, at least try to pinpoint exactly what it is about it that is difficult.

   (b) Given a definition of face, how might you distinguish the infinite face from the others? How do you know that there is only one infinite face?

   (c) Let $T = (V, E)$ be a tree. Try to prove that T is planar. Try to prove that T has only one face.

8. When doing induction on the number of vertices or edges of a graph, we have said that you start with $k + 1$ vertices or edges and *remove* one. It is not equivalent to start with k edges or vertices and *add* one. Why is this so?

9. You have seen that proving that any map can be colored with six colors is straightforward. It is not too much more difficult to prove that all maps can be colored with five. It has long been *empiricially* known by cartographers that *four* colors suffice, but the mathematical question is very difficult and went unsolved for more than one hundred years.

   The four color map problem was reduced to checking more than a thousand possible cases. The question was then "settled" by having a computer check all the cases. Most mathematicians consider the issue resolved, but there is nonetheless some uneasiness about the solution. Historically, mathematical questions have been considered to be solved when experts in the field read the proof and declare it to be sound. The difficulty is that when a computer checks cases and comes forward with an answer, there is no proof to publish and check. Experts can check the computer program that produced the answer, of course, and this was what was done in the case of the four color map theorem. On the other hand, this brings up a host of philosophical questions.

   - Do the experts that check the program have to understand all the workings of the machine that was used to run it? What if there is a problem with the machine? (If you think this is far-fetched, don't forget that the first commercial version of the Pentium computer chip did ordinary division incorrectly!)

   - In the case of the four color theorem, the cases were determined by humans and checked by a machine. What will the mathematical community do when the machine not only checks the cases but also determines what they are in the first place? (Can such a time be very far away?)

   Are there other questions that arise in your mind? What, in your opinion, would be a reasonable course for the mathematical community with respect to computer-generated proofs?[7]

---

[7] For a philosophical discussion along other lines, see *Pi in the Sky* by John D. Barrow (Oxford University Press, 1992.)

5 Functions

5.1 Basic Ideas

In your previous studies you have, no doubt, run across the concept of a function. In fact, you have probably spent a large percentage of your mathematical education considering functions in one context or another. They are in every mathematician's tool chest; we use them all the time. This chapter will formalize the idea of a function and begin to establish some of the accompanying standard ideas and techniques available to us.

You probably first encountered functions in the context of a function from $\mathbb{R}$ to $\mathbb{R}$, that is, a correspondence that takes each real number and assigns to it a real number. There are countless possibilities. The function $f(x) = 3x - 1$ assigns to a given input number the number obtained by multiplying the input by 3 and subtracting 1 from the result. Another function may interpret a given number as a fixed time and assign to it the value of a particular stock at that time. Of course, we need not restrict ourselves to thinking of functions in which the inputs and outputs are numbers. A table in which the first column lists the names of a bunch of dairy cows and the second column indicates the average daily production of milk by each cow is an example of a function.

Name	Milk Produced per Day (in pounds)	Name	Milk Produced per Day (in pounds)
Alice	39.5	Kicker	37
Big Mama	43	Sassy	38.5
Bluebell	35	Spots	38
Clara	39.5	Sue Ellen	39
Jane	40.5	Tilly	39

Or suppose we have a collection of subsets of $\mathbb{N}$. We could assign to each the subset of $\mathbb{N}$ that we get by adding 1 to each element of the original set. For instance,

$$\{1, 5, 10\} \rightarrow \{2, 6, 11\}.$$

In general, a function from a set A to a set B is a correspondence by which an element of B is assigned to every element of A. The essential ingredient here is that *each* element of A have a *single* element of B assigned to it.

In practice, people usually think of functions in this informal way. It conjures up a dynamic picture of something that takes elements of A and *transforms* them into elements of B. Unfortunately, it is not mathematically rigorous. Just what exactly is a "correspondence"? What does it mean to "assign" an element of B to a given element of A? Well, first of all, we are considering *pairs* of elements—each element of A is paired with an element of B. That is, a function is a relation. However, not just any relation (bunch of ordered pairs) will do. We said that the key ingredient in our discussion was the specification that each element of A has a single element of B assigned to it. The following definition lays this out formally.

5.1.1 DEFINITION

Let A and B be nonempty sets. A **function** f from set A to set B (denoted by $f : A \rightarrow B$) is a relation between A and B satisfying the following conditions:

1. For each $a \in A$ there exists $b \in B$ such that $(a, b) \in f$, and

2. if (a, b) and (a, c) are in f, then $b = c$.

If $a \in A$, the unique element $b \in B$ for which $(a, b) \in f$ is denoted by $f(a)$.

Remark.   Using the familiar language of functions, this says that if $a \in A$, we never have to justify the existence of $f(a)$; it is always there. Furthermore, a maps to only one thing. If we know that a maps to b and a also maps to c, then c and b must be equal.

5.1.2 EXERCISE

Let

$$A = \{\diamondsuit, \clubsuit, \heartsuit, \spadesuit\},$$
$$B = \{\flat, \sharp, \natural\},$$
$$C = \{\emptyset, \exists, \forall, \infty, \otimes\}.$$

Which of the following relations are functions from one of the above sets to another? Explain.

1. $\{(\flat, \heartsuit), (\sharp, \clubsuit), (\natural, \spadesuit)\}$
2. $\{(\emptyset, \diamondsuit), (\exists, \spadesuit), (\otimes, \heartsuit), (\infty, \heartsuit)\}$
3. $\{(\flat, \heartsuit), (\sharp, \heartsuit), (\natural, \heartsuit)\}$
4. $\{(\diamondsuit, \flat), (\clubsuit, \flat), (\heartsuit, \sharp), (\spadesuit, \natural)\}$

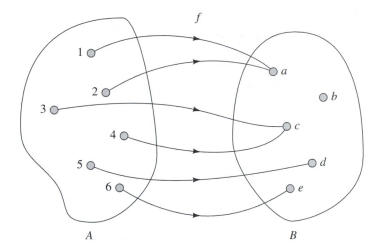

Figure 5.1 This is a useful way to represent a function between two finite sets A and B.

5. $\{(\diamondsuit, \heartsuit), (\clubsuit, \diamondsuit), (\heartsuit, \diamondsuit), (\spadesuit, \heartsuit)\}$
6. $\{(\exists, \diamondsuit), (\varnothing, \clubsuit), (\forall, \heartsuit), (\exists, \heartsuit), (\infty, \diamondsuit), (\otimes, \clubsuit)\}$

Choose one set that you decided was a function. Draw and carefully label a graph of this function as illustrated in Figure 5.1. □

5.1.3 EXERCISE

Consider the function shown in Figure 5.1. What is $f(1)$? What is $f(2)$? What is $f(5)$? □

5.1.4 EXERCISE

When discussing functions from $\mathbb{R}$ to $\mathbb{R}$, many high school teachers tell their students that they can tell from a graph whether they have a function by determining whether the graph passes "the vertical line test"—that is, whether each vertical line in the plane crosses the graph exactly once. Use the language of ordered pairs to explain why this works. □

5.1.5 DEFINITION

In Definition 5.1.1, the set A is called the **domain** of the function f. It is denoted by $\mathcal{D}om(f)$.

The set B is called the **codomain** of the function f. It is denoted by $\mathcal{C}odom(f)$.

The set $\mathcal{R}an(f) = \{b \in B : \text{there exists } a \in A \text{ such that } b = f(a)\}$ is called the **range** of the function f or the **image of** A under the function f.

5.1.6 EXERCISE

Consider the function shown in Figure 5.1. What are the domain, codomain, and range for this function?   □

Two functions are equal if they have the same domain, the same codomain, and "agree" on every element of the domain. More formally,

5.1.7 THEOREM (Equality of functions)

If $f : A \rightarrow B$ and $g : A \rightarrow B$ are functions, then $f = g$ if and only if $f(a) = g(a)$ for all $a \in A$.

(*Hint:* A function is a relation. Saying that two functions are equal is to say they are relations between the same pair of sets, and they include the same ordered pairs.)   □

The tool by which we as mathematicians compare mathematical structures is the function. To make the comparisons meaningful, it is usually desirable to consider functions with special properties. For instance, in the study of calculus, one looks at continuous functions and differentiable functions. In many mathematical contexts it is useful to consider functions that are *one-to-one* and functions that are *onto*.

To get the general idea of why we are interested in these properties, recall our short discussion of isomorphic partial orders on page 73. We said that if two partial orderings were "the same up to a relabeling of the elements," then they were to be called isomorphic partial orders. Of course, we have a good intuitive idea of what it means in that context to relabel elements. Each element in the first partial order can be "associated" with an element in the second partial order in a natural way. This is simply a function from the first set of labels to the second set of labels! However, the word "relabel" has some specific connotations. First of all, we will require that there are enough new names to "go around"—that is, that we won't have to use the same name twice. Each element in the second list can be assigned *only once*. Furthermore, since we are comparing two different structures, we want to account for all the elements in each—we don't want to have any new names left over.

5.1.8 DEFINITION

A function $f : A \rightarrow B$ is said to be **one-to-one** if given $b \in B$, there is at most one $a \in A$ for which $b = f(a)$.

A function $f : A \rightarrow B$ is said to be **onto** if for each $b \in B$, there is at least one $a \in A$ for which $b = f(a)$. In other words, f is onto if the codomain and the range of f are the same set.

A function that is both one-to-one and onto is often called a **one-to-one correspondence**.[1]

---

[1] Other common usage: One-to-one functions are sometimes called **injective**. Onto functions are sometimes called **surjective**. In this spirit, one-to-one correspondences are called **bijective**. Interestingly, many mathematicians mix the usage and say one-to-one, onto, and bijective.

5.1.9   EXERCISE

In the context of relabeling, which of the conditions above corresponds to "having enough new names to go around"? Which corresponds to "not having any new names left over"?

□

5.1.10   EXERCISE

1. Give an example of finite sets A and B, and a function $f : A \to B$ in which f is one-to-one but not onto.

2. Give an example of finite sets A and B, and a function $f : A \to B$ in which f is onto but not one-to-one.

3. Give an example of finite sets A and B, and a function $f : A \to B$ in which f is both one-to-one and onto.

4. Give an example of finite sets A and B, and a function $f : A \to B$ in which f is neither one-to-one nor onto.

Pictures such as Figure 5.1 will help you to visualize the concepts of one-to-one and onto for functions between finite sets.

□

5.1.11   EXERCISE

Thinking some more about "the vertical line test," we can ask about a possible "horizontal line test." Such tests can be used to determine whether the graph of a function from $\mathbb{R}$ to $\mathbb{R}$ depicts a function that is one-to-one, onto, neither, or both. Devise a horizontal line test and use the language of ordered pairs to explain how it works.

□

5.1.12   EXERCISE

1. Give an example of a function $f : \mathbb{R} \to \mathbb{R}$ in which f is one-to-one but not onto.

2. Give an example of a function $f : \mathbb{R} \to \mathbb{R}$ in which f is onto but not one-to-one.

3. Give an example of a function $f : \mathbb{R} \to \mathbb{R}$ in which f is both one-to-one and onto.

4. Give an example of a function $f : \mathbb{R} \to \mathbb{R}$ in which f is neither one-to-one nor onto.

Illustrate your examples by drawing the graphs of the functions that you chose.   □

5.1.13   THEOREM

Let $f : A \to B$ be a function. The following conditions on f are equivalent:

   i. f is one-to-one.

   ii. For all a_1 and a_2 in A, if $f(a_1) = f(a_2)$, then $a_1 = a_2$.

(*Hint:* Try proof by contraposition.)   □

Showing a function is one-to-one: Remember that equivalent statements are just different ways of saying the same thing. Thus the second statement in Theorem 5.1.13 is *another way of defining what it means for a function to be one-to-one.* When attempting to prove that a function $f : A \to B$ is one-to-one, we almost always use the alternative phrasing rather than Definition 5.1.8. This gives us a straightforward procedure for proving that a function is one-to-one:

> *We begin by assuming that $f(a_1) = f(a_2)$ and then show that a_1 must equal a_2.*

(Notice that this is very much like the procedure for showing that something is unique. Explain why this makes sense.)

5.1.14 EXERCISE

For each of the functions $f : \mathbb{R} \to \mathbb{R}$ given below either show that f is one-to-one or prove that it is not.

  1. $f(x) = \frac{x}{2} + 6$ 2. $f(x) = 4x^3$ 3. $f(x) = x^3 - x$
  4. $f(x) = e^x$ 5. $f(x) = \sin(x)$ □

Showing that a function is onto: Stating that a function f is onto is to state an existence theorem: Given an arbitrary element b in the codomain of f, *there exists* an element a in the domain of f such that $f(a) = b$. (See page 28 for a review of the procedure for proving existence theorems.)

5.1.15 EXERCISE

Determine whether the functions given in Exercise 5.1.14 are onto. Prove your answers.

□

We finish the section with a theorem that will be used in Chapter 7.

5.1.16 THEOREM

Let X and Y be sets. If $f : X \to Y$ is a function, then there is an onto function $f^* : X \to \mathcal{C}odom(f)$ such that for all $x \in X$, $f(x) = f^*(x)$.

   In addition, if f is one-to-one, f^* is a one-to-one correspondence.   □

What Theorem 5.1.16 tells us is that if we have a function that is not onto, we can get our hands on a closely related function that *is* onto by changing our point of view a bit—by thinking of the same set of pairs as a function with the same domain and a different (smaller) codomain.

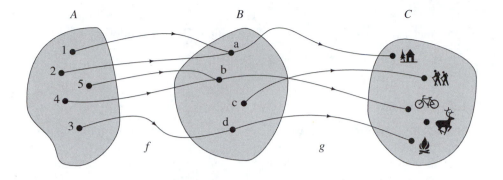

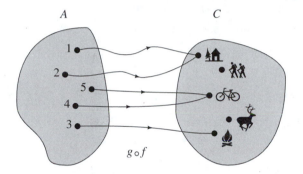

Figure 5.2   Composition of functions

5.2   Composition and Inverses

5.2.1   DEFINITION

If $f : A \to B$ and $g : B \to C$ are functions, then a new function $g \circ f : A \to C$ can be defined by $(g \circ f)(a) = g(f(a))$. This new function is called the **composition** of g and f. (See Figure 5.2.)

5.2.2   EXERCISE

1. Give an example of finite sets A, B, and C, and functions $f : A \to B$ and $g : B \to C$ for which f is onto, but $g \circ f$ is not.

2. Give an example of finite sets A, B, and C, and functions $f : A \to B$ and $g : B \to C$ for which g is onto, but $g \circ f$ is not.

3. Give an example of finite sets A, B, and C, and functions $f : A \to B$ and $g : B \to C$ for which f is one-to-one, but $g \circ f$ is not.

4. Give an example of finite sets A, B, and C, and functions $f : A \to B$ and $g : B \to C$ for which g is one-to-one, but $g \circ f$ is not.   □

5.2.3 THEOREM

Suppose that $f : A \to B$ and $g : B \to C$ are functions. Then the following hold.

1. If f and g are both one-to-one, $g \circ f$ is one-to-one.
2. If f and g are both onto, $g \circ f$ is onto.   □

5.2.4 PROBLEM

Suppose that $f : A \to B$ and $g : B \to C$ are functions. In each of the following cases, answer the question. If your answer is "yes," give a proof. If your answer is "no," give a counterexample and say what additional hypotheses are needed to make the statement true; then prove the statement with the additional hypotheses.

1. If $g \circ f$ is one-to-one, must f be one-to-one?
2. If $g \circ f$ is one-to-one, must g be one-to-one?
3. If $g \circ f$ is onto, must f be onto?
4. If $g \circ f$ is onto, must g be onto?   □

5.2.5 THEOREM

Composition of functions is associative. That is, if $f : A \to B$, $g : B \to C$, and $h : C \to D$ are functions, then

$$h \circ (g \circ f) = (h \circ g) \circ f.$$

(Problem 5.2.6 is a guide for proving this theorem.)   □

5.2.6 PROBLEM

This problem relates to the proof of Theorem 5.2.5. In proving that theorem, you should start, of course, by verifying that

$$h \circ (g \circ f) \quad \text{and} \quad (h \circ g) \circ f$$

have the same domain and codomain. (Do that now!)
    Then it is necessary to show that for all $a \in A$,

$$(h \circ (g \circ f))(a) = ((h \circ g) \circ f)(a).$$

In the table on the next page you will find four arguments purporting to show this fact. Only one is completely correct. The others vary in how far they stray. Critique each argument, describing as many errors as you find. Finally, identify the correct argument.
    (*Note:* The arguments may fall short for a variety of reasons. They may be ill-conceived, starting and ending in the wrong places; they may contain lapses in logic;[2] or they may apply the notion of function composition incorrectly.)

---

[2] Remember the caution made in the introductory essay: "An argument consisting entirely of true statements is not valid if the connecting inferences are not justified by logic."

Argument 1	Argument 2
$(h \circ (g \circ f))(a) = ((h \circ g) \circ f)(a)$	$(h \circ (g \circ f))(a) = h((g \circ f)(a))$
$h \circ (g(f(a))) = (h \circ g) \circ f(a)$	$= h(g(f(a)))$
$h(g(f(a))) = h(g(f(a)))$	$= (h \circ g)(f(a))$
	$= ((h \circ g) \circ f)(a)$

Argument 3	Argument 4
$(h \circ (g \circ f))(a) = h((g \circ f)(a))$	$(h \circ (g \circ f))(a) = ((h \circ g) \circ f)(a)$
$= h(g(f(a)))$	$h((g \circ f)(a)) = (h \circ g)(f(a))$
and	$h(g(f(a))) = h(g(f(a)))$
$((h \circ g) \circ f)(a) = h \circ (g(f(a)))$	
$= h(g(f(a)))$	

□

5.2.7 THEOREM

Let $f : A \rightarrow B$ be a function. The relation

$$I = \{(f(a), a) : a \in A\}$$

is a function if and only if f is a one-to-one correspondence. □

5.2.8 DEFINITION

Let f be a one-to-one correspondence. The function defined in Theorem 5.2.7 is called the **inverse** of the function f. It is denoted by f^{-1}.

5.2.9 THEOREM

Let $f : A \rightarrow B$ be a function.

Part I. Then if f is a one-to-one correspondence, the following statements hold.
  1. $(f \circ f^{-1})(x) = x$ for all $x \in B$.
  2. $(f^{-1} \circ f)(x) = x$ for all $x \in A$.
  3. If $g : B \rightarrow A$ is any function for which
    - $(f \circ g)(x) = x$ for all $x \in B$, and
    - $(g \circ f)(x) = x$ for all $x \in A$,
    then $g = f^{-1}$.

(In other words, f^{-1} satisfies these two composition conditions, and, furthermore, it is the only function that does.)

Part II. Conversely, if f is any function from A to B and there exists a function $g : B \rightarrow A$ satisfying $(f \circ g)(x) = x$ for all $x \in B$ and $(g \circ f)(x) = x$ for all $x \in A$, then f is a one-to-one correspondence. (Thus f^{-1} exists and must, in turn, be equal to g by Part I of the theorem.)   □

This theorem allows us easily to prove the relationship between the inverse of a composition of two functions and the inverses of the functions being composed.

5.2.10 THEOREM

Suppose that $h : A \rightarrow B$ and $k : B \rightarrow C$ are functions. If h and k are both one-to-one and onto, $(k \circ h)^{-1}$ is a function and $(k \circ h)^{-1} = h^{-1} \circ k^{-1}$.   □

5.3   Images and Inverse Images

Theorems 5.2.7 and 5.2.9 show that unless f is one-to-one and onto, there is no meaningful way to define an inverse function for f. It is often useful to consider a more general concept which is closely related to that of the inverse function—the inverse image.

5.3.1 DEFINITION

Let $f : A \rightarrow B$ be a function. Let $S \subseteq B$. The set

$$f^{-1}(S) = \{a \in A : f(a) \in S\}$$

is called the **inverse image** of the set S under the function f.

5.3.2 EXAMPLE

Consider the function illustrated in Figure 5.3. What is $f^{-1}(S)$?   ■

5.3.3 EXERCISE (Say it in words)

Let $f : A \rightarrow B$ be a function. Let $X \subseteq B$. Let $z \in A$. Complete the following sentence:

$$z \in f^{-1}(X) \text{ means that } f(z) \underline{\hspace{3cm}}.$$ □

We have said that the inverse of a function and inverse image of a set under a function are closely related concepts; thus we use similar notation to denote them. Be careful, though. They are not the *same* concept! The inverse of a function is a *function*, and we have seen that this exists only if the function is one-to-one and onto. The inverse image of a set under a function is a *set* and exists for all functions, not just those that are one-to-one correspondences. Get straight both the differences and the similarities, and when you see the symbol f^{-1} be sure that you understand whether it denotes a function or a set. (It should be clear from the context.) Keep the distinction in mind when you do the remaining exercises in this section.

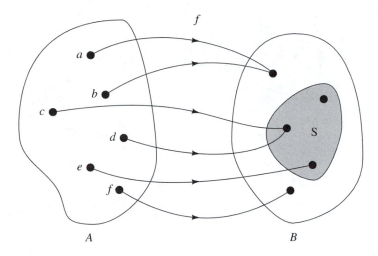

Figure 5.3   What is the inverse image of S?

5.3.4   EXERCISE

Give an example of a function $f : A \rightarrow B$ and a nonempty subset S of B for which $f^{-1}(S) = \emptyset$.   □

5.3.5   EXERCISE

Let $f : \mathbb{R} \rightarrow \mathbb{R}$ be given by $f(x) = x^2$. Find the following.

1. $f^{-1}(\{4\})$
2. $f^{-1}([-2, 9])$
3. $f^{-1}((1, 4])$ □

The following theorem gives some useful properties of inverse images.

5.3.6   THEOREM

Let $f : A \rightarrow B$ be a function. Let Λ be an arbitrary indexing set. Let $\{S_\alpha\}_{\alpha \in \Lambda}$ be a collection of subsets of B, and let S be any subset of B. Then

1. $f^{-1}\left(\bigcup_{\alpha \in \Lambda} S_\alpha\right) = \bigcup_{\alpha \in \Lambda} f^{-1}(S_\alpha)$.
2. $f^{-1}\left(\bigcap_{\alpha \in \Lambda} S_\alpha\right) = \bigcap_{\alpha \in \Lambda} f^{-1}(S_\alpha)$.
3. $f^{-1}(S^{\complement}) = (f^{-1}(S))^{\complement}$.

(*Hint:* Since these are sets, you will need to employ element arguments.)   □

In keeping with the spirit and notation of Definition 5.3.1, we define the image of a subset of A under the function f.

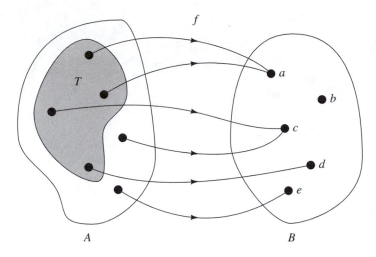

Figure 5.4   What is the image of T?

5.3.7 DEFINITION

Let $f : A \rightarrow B$ be a function. Let $T \subseteq A$. The set

$$f(T) = \{b \in B : \text{ there is some } t \in T \text{ with } f(t) = b\}$$
$$= \{f(t) : t \in T\}$$

is called the **image** of T under the function f.

5.3.8 EXERCISE

Consider the function illustrated in Figure 5.4. What is $f(T)$?   □

5.3.9 EXERCISE (Say it in words.)

Let $f : A \rightarrow B$ be a function. Let $Y \subseteq A$. Let $c \in B$. Complete the following sentence:

$c \in f(Y)$, provided that there exists _____.   □

5.3.10 EXERCISE

Let $f : \mathbb{R} \rightarrow \mathbb{R}$ be given by $f(x) = x^2$. Find $f([-2, 10])$.   □

5.3.11 THEOREM

Let $f : A \rightarrow B$ be a function. Let Λ be an arbitrary indexing set. Let $\{T_\alpha\}_{\alpha \in \Lambda}$ be a collection of subsets of A. Then

   1. $f\left(\bigcup_{\alpha \in \Lambda} T_\alpha \right) = \bigcup_{\alpha \in \Lambda} f(T_\alpha)$.

   2. $f\left(\bigcap_{\alpha \in \Lambda} T_\alpha \right) \subseteq \bigcap_{\alpha \in \Lambda} f(T_\alpha)$.   □

(Remember that there is no function f^{-1}. If you invoked such a function anywhere in your proof of Theorem 5.3.11, go back and rethink!)

5.3.12   PROBLEM

Let $f : A \to B$ be a function.

1. Give an example of a function $f : A \to B$ and two subsets X and Y of A such that $f(X \cap Y) \neq f(X) \cap f(Y)$.
2. Show that $f(\bigcap_{\alpha \in \Lambda} T_\alpha) = \bigcap_{\alpha \in \Lambda} f(T_\alpha)$ for all choices of $\{T_\alpha\}_{\alpha \in \Lambda}$ if and only if f is one-to-one.

   (*Hint:* Prove ($\Longrightarrow$) by contrapositive. Be careful not to ignore the quantification over the collections of sets. It is *very* crucial.)   □

5.4   Order Isomorphisms

We are finally prepared to define isomorphism of partial orderings carefully. I said that two partially ordered sets are to be considered isomorphic if their order structure is "the same up to a relabeling of the elements." As I tried to motivate the definition for one-to-one and onto functions, I hinted that this relabeling was simply a one-to-one correspondence between the underlying sets. Of course, as we relabel the elements, we must preserve the ordering of the elements. This motivates a two-part definition.

5.4.1   DEFINITION (Order-preserving function)

Let $(A, \leq_A)$ and $(B, \leq_B)$ be partially ordered sets. A function $f : A \to B$ is said to be **order-preserving** if for all $x, y \in A$,

> I use notation $\leq_A$ and $\leq_B$ to emphasize that the orderings on A and B are very likely different.

$$x \leq_A y \text{ implies } f(x) \leq_B f(y).$$

5.4.2   THEOREM

Let $(A, \leq_A)$ be a totally ordered set. Let $(B, \leq_B)$ be a partially ordered set. Let $f : A \to B$ be an onto, order-preserving function. Then $(B, \leq_B)$ is a totally ordered set.   □

5.4.3   DEFINITION (Order isomorphism)

Let $(A, \leq_A)$ and $(B, \leq_B)$ be partially ordered sets. A function $f : A \to B$ is called an **order isomorphism** if f is one-to-one, onto, and order-preserving and f^{-1} is order-preserving, also.

Two partially ordered sets are said to be **order isomorphic** if there is an order isomorphism between them.

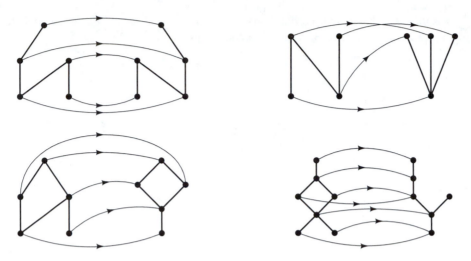

Figure 5.5   Functions between partially ordered sets

5.4.4   EXERCISE

Look at the functions between partially ordered sets depicted in Figure 5.5. For each one, decide whether it is order-preserving, and, if so, whether it is an order isomorphism. Justify your answers.   □

5.4.5   THEOREM

Let $(A, \leq_A)$ and $(B, \leq_B)$ be partially ordered sets. Let $f : A \to B$ be a one-to-one correspondence. Prove that the following statements are equivalent.

   i. f is an order isomorphism

   ii. For all $x, y \in A$, $x \leq_A y$ if and only if $f(x) \leq_B f(y)$.   □

5.4.6   EXERCISE

Show that $(\mathbb{N}, \leq)$ is order isomorphic to $(2\mathbb{N}, \leq)$. (Where $2\mathbb{N}$ is the set of even natural numbers.)   □

5.4.7   THEOREM

Let $(A, \leq_A)$ and $(B, \leq_B)$ be order isomorphic partially ordered sets. Then A is totally ordered if and only if B is totally ordered.   □

> "Let $(A, \leq_A)$ and $(B, \leq_B)$ be order isomorphic partially ordered sets," guarantees the existence of an order isomorphism. Your immediate response should be to say: "Let $f : A \to B$ be an order isomorphism." This will give you something to work with.

   The study of isomorphism is essentially a study of the preservation of mathematical structure. Our intuitive discussion of order isomorphism said that order isomorphic partially ordered sets should have the same order structure. Theorem 5.4.7 gave us our first rigorous view of this idea. The next few theorems and several problems at the end of the chapter continue to explore the preservation of order structures through order isomorphism.

5.4.8   THEOREM

Let $(A, \leq_A)$ and $(B, \leq_B)$ be partially ordered sets. Suppose that $f : A \to B$ is an order isomorphism. Let x and $y \in A$. Then

1. $x = y$ if and only if $f(x) = f(y)$.
2. x and y are unrelated if and only if $f(x)$ and $f(y)$ are unrelated.
3. $x < y$ if and only if $f(x) < f(y)$.   □

5.4.9   THEOREM

Let A and B be order isomorphic partially ordered sets. An element $x \in A$ is a maximal (respectively minimal) element of A if and only if $f(x)$ is a maximal (respectively minimal) element of B.   □

5.4.10   PROBLEM

Let

$$A = \{x \in \mathbb{R} : x = 1 - \frac{1}{n} \text{ for some } n \in \mathbb{N}\},$$

and let $B = A \cup \{1\}$. Consider each of these as a totally ordered set under the usual ordering on $\mathbb{R}$. Show that A is order isomorphic to $\mathbb{N}$ and that A is *not* order isomorphic to B.[3]

(*Hint:* When trying to prove that two partially ordered sets are *not* order isomorphic, it is often useful to try proof by contradiction.)   □

> **Some food for thought:** In Problem 5.4.10, it is tempting to say that *A* and *B* fail to be order isomorphic because there is no one-to-one correspondence between them. Avoid this trap. There are *many* one-to-one correspondences between *A* and *B*, they just don't happen to be order-preserving. (*Can you find a one-to-one correspondence between A and B?*)

---

[3] Problem 5.4.10 hints at the ideas of ordinality and ordinals. Ordinality is closely related to cardinality, which you will study in Chapter 7. It is a way of trying to get at the concept of an infinite number. If you are intrigued, ask a mathematician to recommend a good book on set theory.

5.4.11 THEOREM

Let $(A, \leq_A)$ and $(B, \leq_B)$ be isomorphic partially ordered sets. Let K be a subset of A, let $x \in A$, and let $f : A \to B$ be an order isomorphism.

1. x is an upper bound for K in A if and only if $f(x)$ is an upper bound for $f(K)$ in B.

2. x is the least upper bound for K in A if and only if $f(x)$ is the least upper bound for $f(K)$ in B.

Parallel statements hold for lower bounds and greatest lower bounds. (You should formulate these statements and verify that your arguments for upper and least upper bounds can be adapted to prove them.) □

5.5 Sequences

One of the most central notions in mathematics is that of the sequence. Loosely speaking, a sequence in a set A is an infinite "string" of elements of A, with one element being designated as the 1st term of the sequence, another (not necessarily distinct from the 1st) as the 2nd term of the sequence, and so on. The following are sequences of real numbers.

- $1, 0, 1, 0, 0, 1, 0, 0, 0, 1, \ldots$
- $3, 6, 2, 9, 14, 16, 10, 100, 99, 23, 15, 1003, \ldots$
- $1, \frac{1}{2}, \frac{1}{3}, \frac{1}{4}, \frac{1}{5}, \ldots$

The last of these examples has 1 as its 1st term, $\frac{1}{2}$ as its 2nd term, and so on; in general, $\frac{1}{n}$ is called the nth term of the sequence. That is, with each natural number n we are associating a real number, $\frac{1}{n}$. This discussion should remind you of the concept of a function. We can now give a mathematical definition for a sequence.

5.5.1 DEFINITION

Let A be a nonempty set. Any function $s : \mathbb{N} \to A$ is called a **sequence in** A.

When we think of a function, we rarely picture a set of ordered pairs. When we consider a sequence s in A, we rarely think explicitly of the function $s : \mathbb{N} \to A$. Instead, we keep in mind only the order of the terms. If we were to follow the ordinary notation for functions, we would refer to the terms of a sequence s by $s(1), s(2), \ldots$; however, we usually use different notation for sequences.

SEQUENCE NOTATION: If s is a sequence in A, we refer to the 1st term in the sequence as s_1, the 2nd in the sequence as s_2, the 45th term in the sequence as s_{45}, and so forth. So the notation s_n encodes two pieces of information:

- s_n is an element of the set A and indicates the *value* of the term.
- n is a natural number and indicates the *position* of the term in the sequence.

Our shorthand for the sequence as a whole is (s_n).

5.5.2 EXERCISE

Let (s_n) be the sequence in $\mathbb{R} \times \mathbb{R}$ given by

$$s_n = \left((-1)^n \frac{n}{n+1}, 2n+1 \right).$$

Plot the first seven terms of this sequence in the Cartesian plane. □

Many common examples of sequences are defined *recursively*. The first one or more terms are given, and a rule is specified for calculating the nth term from some or all of the first $n - 1$ terms.

5.5.3 EXAMPLE (The Fibonacci sequence)

We construct a sequence f of natural numbers as follows. Let $f_1 = 1$ and $f_2 = 1$. For all $n > 2$ define

$$f_n = f_{n-1} + f_{n-2}.$$

Find the first several terms of the Fibonacci sequence. □

5.5.4 EXERCISE

Give a recursive definition for the factorial function $n! = 1 \cdot 2 \cdot 3 \cdots n$. □

A special example of a recursively defined sequence is an iterated map.

5.5.5 DEFINITION

Let A be a set, and let $s_0 \in A$. Let $f : A \to A$ be a function. For each $i \in \mathbb{N}$, define:

$$s_i = f(s_{i-1}).$$

The resulting sequence in A is called the **iterated map on f based at s_0**.

> By tradition, the numbering of iterated maps usually starts with 0 rather than with 1. Thus we have $s_0, s_1, s_2, \ldots$ rather than $s_1, s_2, s_3, \ldots$.

5.5.6 EXERCISE

Consider the function $f : \mathbb{R} \to \mathbb{R}$ given by $f(x) = x^2$. Find the first several terms of each of four iterated maps on f based at $s_0 = 0$, $s_0 = 2$, $s_0 = -1$, and $s_0 = -\frac{1}{3}$. □

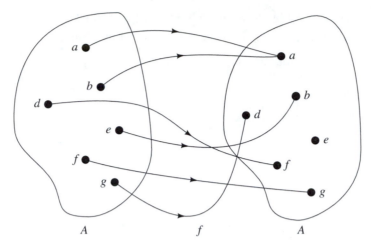

Figure 5.6 $f : A \to A$

5.5.7   EXERCISE

Consider the function shown in Figure 5.6. Find *all* of the iterated maps on f.   □

Mathematical induction is particularly well suited to proving facts about sequences. Use induction to prove the following (very easy) theorem about iterated maps.

5.5.8   THEOREM

Let $f : A \to A$ be a function. Let (s_n) be the sequence in A that is generated by iterating f at s_0. If for some $n \in \mathbb{N}$, $s_{n+1} = s_n$, then $s_m = s_n$ for all $m \geq n$.   □

Induction can be used to prove some more interesting (and less obvious) things, too. Use induction to prove the following facts about the Fibonacci sequence.

> Sequences that behave as the one described in Theorem 5.5.8 are said to be *eventually constant*. Which of the iterated maps in Exercise 5.5.6 is eventually constant? What about the maps in Exercise 5.5.7?

5.5.9   THEOREM

Let (f_n) be the Fibonacci sequence. Then (f_n) has the following properties:

1. For all $n \geq 2$, $f_{n-1} \leq f_n \leq 2 f_{n-1}$.
2. For all $n \in \mathbb{N}$,

$$\sum_{j=1}^{n} f_j = f_{n+2} - 1.$$

3. Let $\phi_1 = \frac{1+\sqrt{5}}{2}$ and $\phi_2 = \frac{1-\sqrt{5}}{2}$. Then for each $n \in \mathbb{N}$,

$$f_n = \frac{1}{\sqrt{5}}(\phi_1^n - \phi_2^n).$$

(*Hint:* First show that $\phi_1 = -1/\phi_2$ and $\phi_1 + \phi_2 = 1$. You will need these identities in the induction step.) □

Sequences with Special Properties

5.5.10 DEFINITION

Let A be a set. Let (s_i) be a sequence in A.

1. (s_i) is a **sequence of distinct terms** if $s_i \neq s_j$ for distinct $i, j \in \mathbb{N}$.
2. (s_i) is a **constant sequence** if there exists $a \in A$ such that $s_i = a$ for all $i \in \mathbb{N}$.

5.5.11 EXERCISE

Give examples of a constant sequence, a sequence of distinct terms, and a sequence that is neither constant nor of distinct terms. □

We frequently will consider sequences in ordered sets. Sequences in ordered sets may have some special properties that relate to the ordering.

5.5.12 DEFINITION

Suppose that A is a totally ordered set.[4] Let (s_i) be a sequence in A.

1. (s_i) is said to be **increasing** if whenever $n \leq m$, $s_n \leq s_m$. (s_i) is **strictly increasing** if whenever $n < m$, $s_n < s_m$.
2. (s_i) is said to be **decreasing** if whenever $n \leq m$, $s_n \geq s_m$. (s_i) is **strictly decreasing** if whenever $n < m$, $s_n > s_m$.
3. (s_i) is said to be **monotonic** if it is either increasing or decreasing.
4. (s_i) is said to be **bounded from below** if there exists $b \in A$ such that for all $i \in \mathbb{N}$, $b \leq s_i$. In this case, b is called a **lower bound** for (s_i).
5. If there exists $c \in A$ such that for all $i \in \mathbb{N}$, $s_i \leq c$, then we say that (s_i) is **bounded from above**. In this case, c is called an **upper bound** for (s_i).
6. If (s_i) is bounded from above and from below, we say that it is **bounded**.

---

[4] Actually, these definitions work just as well in partially ordered sets, but the ideas are most commonly applied to totally ordered sets (like $\mathbb{R}$, for instance).

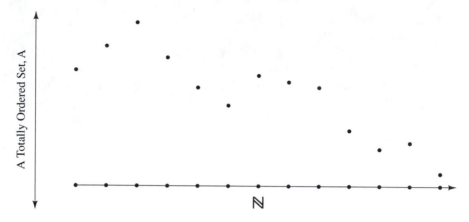

Figure 5.7 This graph shows the way a sequence goes "up and down" in the totally ordered set A (shown on the vertical axis). $\mathbb{N}$, on the horizontal axis, indicates the order of the terms.

5.5.13   EXERCISE

The sort of diagram shown in Figure 5.7 is useful for picturing sequences in totally ordered sets. Use a similar diagram to illustrate each of the terms given in Definition 5.5.12.

□

5.5.14   EXERCISE

Construct a sequence of real numbers that is:

1. strictly increasing.
2. decreasing, but neither strictly decreasing nor constant.
3. a sequence of distinct terms but is not monotonic.
4. bounded from below, but not from above. *(Can you give a nonincreasing example?)*
5. bounded neither from above nor from below.
6. a bounded sequence of distinct terms.

□

5.5.15   THEOREM

Let (n_i) be a strictly increasing sequence of natural numbers. Prove that for all $i \in \mathbb{N}$, $i \le n_i$.

□

5.5.16   EXERCISE

What does Theorem 5.5.15 say about the relationship between the position and the value of a term in a strictly increasing sequence of natural numbers?

□

Subsequences

Sometimes when we have a sequence, we wish to construct a new sequence containing only some of the original terms.

5.5.17 EXAMPLE

Suppose that we start with a sequence of perfect squares.

$$(s_i) = 0, 1, 4, 9, 16, 25, 36, 49, 64, 81, \ldots.$$

We might want to "extract" from (s_i) the sequence consisting only of the terms of (s_i) that are also odd integers:

$$1, 9, 25, 49, 81, \ldots.$$

Notice that the terms of this new sequence are

$$\underbrace{s_2}_{1^{st}}, \underbrace{s_4}_{2^{nd}}, \underbrace{s_6}_{3^{rd}}, \underbrace{s_8}_{4^{th}}, \underbrace{s_{10}}_{5^{th}}, \ldots.$$

That is, the ith term of this sequence is the same as the $2i^{th}$ term of the original sequence. The subscripts on the new sequence form a strictly increasing sequence of natural numbers: $2, 4, 6, 8, \ldots$. The sequence of odd squares is an example of a "subsequence."
■

As always, we need a mathematical definition to drive the mathematics that goes with our intuitive idea.

5.5.18 DEFINITION

Let A be a set. Let (s_i) be a sequence in A. If (n_i) is a strictly increasing sequence in $\mathbb{N}$, Then the sequence

$$s_{n_1}, s_{n_2}, s_{n_3}, \ldots$$

is a **subsequence** of (s_i). We denote this subsequence by (s_{n_i}).

When thinking about subsequences, it is useful to keep in mind that encoded in the notation s_{n_i} are *three* pieces of information:

- s_{n_i} is an element of A and gives the *value* of the term.

- n_i is a natural number and indicates the *position* of the term in the *original sequence*.

- i is also a natural number and indicates the *position* of the term in the *subsequence*.

As a way of making sense of these three pieces of information, think about the difference between the terms $s_{n_{k+1}}$, s_{n_k+1}, and $s_{n_k} + 1$. The exact placement of that "+1" is very important!

5.5.19 EXERCISE

Consider the following sequence (s_i) of letters:

T,H,I,S̶, I,S, T,H̶,E̶, S̶,O,N̶,G, T,H,A,T, D,O̶,E,S,N,T̶, E,N,D,
Y̶,E̶,S, I,T̶, G,O,E,S, O,N, A,N,D̶, O̶,N̶, M,Y, F,R,I,E̶,N,D, . . .

The terms that are boxed represent a subsequence (s_{n_i}) of (s_i).

1. List $s_1, s_2, s_3, \ldots, s_{10}$.
2. List $s_{n_1}, s_{n_2}, s_{n_3}, \ldots, s_{n_{10}}$.
3. What is s_3?
4. What is s_{n_3}?
5. What is n_7?
6. For $k = 6, 7, 8$, find s_{n_k+1} and $s_{n_{k+1}}$. (Notice that $s_{n_k} + 1$ does not even make any sense!)   □

5.5.20 EXERCISE

Explain what Theorem 5.5.15 says about the relationship between the position of a term in a subsequence and its position in the original sequence. (This is meant to be an informal explanation. Phrases like "appears earlier" or "appears later" might be helpful.)   □

"Passing to a subsequence" preserves many important properties of the original sequence. The following (easy to prove) theorems will help you familiarize yourself with subsequence notation.

5.5.21 THEOREM

Let A be a set, and let (s_i) be a sequence in A. Then

1. If (s_i) is constant, every subsequence of (s_i) is constant.
2. If (s_i) has distinct terms, every subsequence of (s_i) has distinct terms.   □

5.5.22 THEOREM

Let A be a totally ordered set, and let (s_i) be a sequence in A. If (s_i) is increasing, every subsequence of (s_i) is increasing. (The statement is still true if the word "increasing" is replaced by "decreasing," "monotonic," "bounded above," "bounded below," or "bounded.")   □

Constructing Subsequences Recursively

Here's a common sort of problem.

Suppose (s_i) is a sequence. Find a subsequence of (s_i) that has such and such a property.

It often happens that the only way to obtain the desired subsequence is to choose the terms one by one, that is, recursively. To do this, we will have to provide a "recipe" that tells us (1) how to choose the first term of the subsequence, and (2) how to choose later terms of the subsequence based on terms already chosen.

Mathematical induction is often necessary to make this process work. It plays a dual role.

- It proves that the recursive definition "works." That is to say, it shows that when we say "choose the $(k + 1)^{st}$ term in this way . . . ," the process we prescribe can actually be carried out.
- It proves that the resulting subsequence has the property we are looking for.

I am going to illustrate these ideas by proving the following theorem for you.

5.5.23   THEOREM

In a totally ordered set, any sequence that does not have a smallest term has a decreasing subsequence.

Before launching into the proof, some preliminary remarks are in order. The intuition here is this: As I construct my decreasing subsequence, I will have to choose terms that get smaller and smaller. Since no term in the original sequence can be the smallest, then at any stage of the subsequence construction there must always be a smaller term available. The tricky thing is that the terms in the subsequence must appear in the same order that they do in the original sequence. If my first subsequence term is s_1 and the second one is s_{12} (chosen so that $s_{12} < s_1$), then I cannot choose s_4 for my third term even if $s_4 < s_{12}$. When I pick the next term for my subsequence, therefore, I will have to take care to pick it *farther out* in the original sequence. In the proof of Theorem 5.5.23, try to spot the part of the proof that takes care of this requirement.

Proof.   Let A be a totally ordered set. Suppose (s_n) is a sequence in A with no smallest term. I will recursively construct a decreasing subsequence of (s_n).

Choose $n_1 = 1$. Then $s_{n_1} = s_1$.

INDUCTION HYPOTHESIS: Suppose we have chosen $n_1, n_2, n_3, \ldots, n_k$ so that the following three properties hold:

1. $n_1 < n_2 < \cdots < n_k$.
2. $s_{n_1} > s_{n_2} > s_{n_3} > \cdots > s_{n_k}$.
3. If $i < n_k$, then $s_i > s_{n_k}$.

Our job is to show that we can choose n_{k+1} so that $n_1, n_2, \ldots, n_k, n_{k+1}$ satisfy the same three properties.

Because s_{n_k} is not the smallest term in the sequence, there exists $j \in \mathbb{N}$ such that $s_j < s_{n_k}$. Choose n_{k+1} to be the smallest such index. (That is, $s_{n_{k+1}}$ is the *first* term in the sequence that is smaller than s_{n_k}.)

Now we check properties 1–3 above for $n_1, n_2, \ldots, n_k, n_{k+1}$.

1. Property 3 in the induction hypothesis tells us that n_{k+1} must be *larger* than n_k. *(Why?)* So

$$n_1 < n_2 < \cdots < n_k < n_{k+1}.$$

2. Our choice of n_{k+1} guarantees that $s_{n_{k+1}} < s_{n_k}$, so

$$s_{n_1} > s_{n_2} > \cdots > s_{n_k} > s_{n_{k+1}}.$$

3. The fact that n_{k+1} is the *smallest* index such that $s_{n_{k+1}} < s_{n_k}$, tells us that for all $i < n_{k+1}, s_i \geq s_{n_k} > s_{n_{k+1}}$.

The resulting sequence is a *sub*sequence of (s_n) by property 1. It is decreasing by property 2.   ■

DECRYPTING THE PROOF: This proof is written in a conventional way that may at first seem arcane and hard to decipher. But in fact, the proof is itself a blueprint for understanding the thought process behind the construction of the subsequence. Examination of the proof reveals four major stages in the proof, which are typical for this sort of argument.

> **General structure:** The construction of a subsequence comes in four basic stages.
>
> 1. *The base case:* State how the 1st term is chosen.
> 2. *The induction hypothesis:* Assume the first k terms have been chosen and specify what properties these are supposed to have.
> 3. *The induction step (part I):* Indicate how the $(k+1)^{st}$ term is to be chosen.
> 4. *The induction step (part II):* Show that if you add the new term to the list of terms already chosen, the updated list will also satisfy the properties laid out in the induction hypothesis. (This guarantees that the process is repeatable.)

Now go back to the proof of Theorem 5.5.23. Start by identifying the four stages within the proof. The actual process that is used for picking the subsequence is spelled out in the base case and in the first part of the induction step. (Ignore the other parts of the proof for the moment.) To help you see how my construction works, you should look at an example. Refer again to the sequence depicted in Figure 5.7 (page 122). Apply my construction to find the first few terms of the subsequence; circle the chosen terms as you go.

Next, you should try to understand what the induction hypothesis is saying. To help you do this, verify that the terms you have chosen from Figure 5.7 satisfy the three properties in the induction hypothesis.

Finally, read through the second part of the induction step. Think carefully about the role of the *third* property. I don't ultimately need it to draw my conclusion, but it is crucial nonetheless. What is it doing?

MAKING YOUR OWN PROOFS: When faced with the problem of constructing a subsequence, you will need to do some preliminary work. You should start by experimenting with an example. In fact, I began work on the proof of Theorem 5.5.23 by thinking about a picture very like Figure 5.7. Studying an example will help you to understand why the theorem is true. It should also suggest the process by which you will choose your subsequence. If one example is not enough, look at others. Once you get an inkling of what needs to be done, try your idea out on a different example to make sure it works.

Armed with the insights you have gleaned from examples, you should return to the general case and carefully write out the process for picking the first few terms of your subsequence. (Each step will probably require a short paragraph.) You will notice several things, which will become elements of your proof.

There is an art to choosing a good, instructive example. It should be simple enough to work with easily, but it should also be rather "generic," in the sense that it should not have special properties that are not guaranteed by the hypothesis of the theorem. For instance, the sequence shown in Figure 5.7 went up and down. If it had been a decreasing sequence, it would not have been very revealing.

- Your successive terms (with the possible exception of the 1st one) are all chosen using exactly the same procedure. *This procedure will turn into the first part of your induction step.*

- The *justification* you give for being able to make your choice is the same regardless of whether you are going from the 2nd term to the 3rd term or from the 12th to the 13th. *This justification will be the second part of your induction step.*

- Your induction hypothesis will include two kinds of things:
  1. the facts you use at each step to justify your procedure for choosing the next term.
  2. the facts necessary so that the subsequence will ultimately satisfy the desired properties.

To write the proof, you will need to translate these observations into the language of sequences and subsequences. This can be a little delicate, and may require several revisions before you have a "clean" proof.

Here are some theorems on which to try your hand. The first one is easy.

5.5.24 THEOREM

Let (s_i) be a sequence of natural numbers. Then (s_i) has a subsequence of even numbers if and only if for all $n \in \mathbb{N}$ there exists $j \geq n$ such that s_j is even.

The statement remains true if you replace the word "even" with the word "odd"—or even the word "prime."

(*Hint:* One direction of this equivalence does not require induction at all. Which is it?) □

5.5.25 THEOREM

Prove that every sequence in a totally ordered set has a monotonic subsequence.

(*Hint:* Use the contrapositive of Theorem 5.5.23.) □

5.5.26 LEMMA

Let (s_i) be a sequence in a set A. Then (s_i) has a constant subsequence if and only if there exists $a \in A$ such that for all $n \in \mathbb{N}$ there exists $j \geq n$ such that $s_j = a$. □

5.5.27 EXERCISE

In the following theorem you will need to be able to say what it means for a sequence *not* to have a constant subsequence. In preparation for that, negate the statement in Lemma 5.5.26. □

5.5.28 THEOREM

Let (s_i) be a sequence in a set A. Then either (s_i) has a constant subsequence or (s_i) has a subsequence of distinct terms. □

5.6 Binary Operations

In Section 5.5, we saw that sequences are really just a special kind of function. There are other everyday mathematical objects that are really functions in disguise. Among them are all the usual arithmetic operations. Addition, subtraction, and multiplication are examples of *binary operations*. When we add two real numbers together, we take two real numbers and we associate with this *pair* a third real number. For instance, with the pair (2, 7) we associate the number 9. This leads us to the following definition.

5.6.1 DEFINITION

Let A be a set. A function from $A \times A$ to A is called a **binary operation** on A.

Remark. In keeping with the usual conventions of arithmetic, for a binary operation $*$ we write $a * b = c$ when the input of the pair (a, b) gives an output of c.

5.6.2 EXAMPLE

Using the language of the definition, explain why the following are binary operations on the specified sets.

    1. Addition, subtraction, and multiplication of real numbers.

2. "Averaging" on $\mathbb{R}$

$$\text{average}(a, b) = \frac{a + b}{2}.$$

3. Exponentiation on $\mathbb{N}$

$$x\char`^y = x^y.$$ ∎

5.6.3   EXERCISE

Explain why division is not a binary operation on $\mathbb{R}$ and why subtraction is not a binary operation on $\mathbb{N}$.   □

5.6.4   PROBLEM

Define $*$ like this: For real numbers a, b, and c, $c = a * b$ if $a^2c^2 = b^2$. Is $*$ a binary operation? Why or why not?   □

   Notice that the domain of a binary operation $*$ on A is the set of *ordered* pairs of elements from the set A. Thus we are making no assumptions about whether $a * b = b * a$ for $a, b \in A$. However, this is true for some operations.

5.6.5   DEFINITION

A binary operation $*$ on a set A is said to be **commutative** if

$$a * b = b * a$$

for all $a, b \in A$.

5.6.6   EXERCISE

Which of the binary operations in Example 5.6.2 are commutative and which are not? Explain.   □

   Binary operations take two elements and yield a third. There are many instances in which we would like to apply an operation to three or more elements and obtain a meaningful result. The total bill at the grocery store, for example, is the sum of the prices of all the individual purchases, which we can calculate even though (strictly speaking) addition is only defined for pairs of numbers.
   Consider for a moment an expression like $a * b * c$. What might this mean? The binary operation $*$ is defined only for two elements. We might therefore proceed by obtaining $d = a * b$ and then finding $d * c$. Following the usual convention in which operations in parentheses are performed first, we would denote this by

$$(a * b) * c.$$

Alternatively, we might calculate $e = b * c$ and then find $a * e = a * (b * c)$. These are the only reasonable possibilities, but there is no a priori reason to prefer one over the other. If $a * b * c$ is to be unambiguously understood, we must have

$$(a * b) * c = a * (b * c).$$

5.6.7 DEFINITION

A binary operation $*$ on a set A is said to be **associative** if $(a * b) * c = a * (b * c)$ for all $a, b, c \in A$.

Since many common algebraic manipulations require repeated application of a binary operation, associative operations play an especially important role in algebra.

5.6.8 EXERCISE

Which of the binary operations in Example 5.6.2 are associative and which are not? Explain.   □

5.6.9 EXERCISE

Let A be any set. Define $a * b = a$ for all $a, b \in A$.

1. Show that $*$ is a binary operation.
2. Is $*$ commutative?
3. Is $*$ associative?   □

5.6.10 EXERCISE

Let S be any set. Show that $\cap$ and $\cup$ are associative and commutative binary operations on $\mathcal{P}(S)$.   □

5.6.11 PROBLEM

Let $(A, \leq)$ be a totally ordered set. Define

$$\max(a, b) = \begin{cases} a & \text{if } b < a, \\ b & \text{if } a \leq b. \end{cases}$$

Show that max is a binary operation on A. Is it associative? Is it commutative?   □

■ PROBLEMS

1. Determine whether each of the following functions is one-to-one, onto, neither, or both. Prove your answers.

   (a) $f : \mathbb{R} \to [1, \infty)$, given by $f(x) = x^2 + 1$.

   (b) $f : (2, \infty) \to (1, \infty)$, given by $f(x) = \frac{x}{x-2}$.

   (c) $f : \mathbb{R} \times \mathbb{R} \to \mathbb{R}$, given by $f(x, y) = x^2 + y^2$.

(d) $f : \mathbb{R} \times \mathbb{R} \to \mathbb{R}$, given by $f(x, y) = x - y$.

(e) $f : \mathbb{N} \to \mathbb{N} \times \mathbb{N}$, given by $f(n) = (n, n)$.

2. (a) Let $a, b \in \mathbb{R}$, with $a \neq 0$. Show that $f : \mathbb{R} \to \mathbb{R}$, given by $f(x) = ax + b$ is a one-to-one correspondence. (What happens if $a = 0$?)

(b) Show that for all positive real numbers k, $f : (-k^2, \infty) \to (-\infty, k)$, given by

$$f(x) = \frac{kx}{x + k^2}$$

is a one-to-one correspondence.

(c) Show that there exists no real number k for which $f : \mathbb{R} \to \mathbb{R}$, given by $f(x) = \sin(kx)$ is one-to-one. Show also that there is no k for which it is onto.

(d) Show that there exists no positive real number k for which $f : \mathbb{R} \to \mathbb{R}$, given by $f(x) = \sqrt{kx^2 + 5}$ is a one-to-one correspondence.

3. Let A and B be sets. Let $S \subseteq A \times B$. We can define functions $\Pi_1 : S \to A$ and $\Pi_2 : S \to B$ by

$$\Pi_1(a, b) = a \quad \text{and} \quad \Pi_2(a, b) = b.$$

Π_1 is called the **projection of S into A** and Π_2 is called the **projection of S into B**.

(a) Show by giving examples that Π_1 need be neither one-to-one nor onto and likewise for Π_2.

(b) Suppose that $S : A \to B$ is a function. (Remember, a function is a set of ordered pairs!) Must it be true that Π_1 is one-to-one? Π_1 is onto? Π_2 is one-to-one? Π_2 is onto? Prove your answers.

4. Let $f : A \to B$ be a function. Define a relation on A by

$$x \sim y \quad \text{if } f(x) = f(y).$$

Prove that $\sim$ is an equivalence relation. Describe the equivalence classes.

5. *"Cancellation laws" for function composition—Part I*

(a) Show that it is possible to find sets A, B, and C and functions $f : B \to C$, $g : A \to B$, and $h : A \to B$ for which $f \circ g = f \circ h$ and yet $g \neq h$. (*Hint:* Draw pictures!)

(b) Let $f : B \to C$ be a function. Complete the following statement and then prove it:

If f is _____, then for all sets A and all functions $g : A \to B$ and $h : A \to B$, $f \circ g = f \circ h$ implies that $g = h$.

(c) Prove the converse of the statement you came up with in part (b). (*Hint:* Proof by contrapositive works best here. The hardest part is negating the complicated statement. Then use the insight you gained in doing part (a).)

6. *"Cancellation laws" for function composition—Part II*

(a) Show that it is possible to find sets A, B, and C and functions $f : A \to B$, $g : B \to C$, and $h : B \to C$ for which $g \circ f = h \circ f$ and yet $g \neq h$. (*Hint:* Draw pictures!)

(b) Let $f : A \to B$ be a function. Complete the following statement and then prove it:

If f is _____, then for all sets C and all functions $g : B \to C$ and $h : B \to C$, $g \circ f = h \circ f$ implies that $g = h$.

(c) Prove the converse of the statement you came up with in part (b). (Same *hint* as in part (c) of Problem 5.)

7. Suppose $f : A \to B$ is a one-to-one correspondence. Show that f^{-1} is also a one-to-one correspondence and that $(f^{-1})^{-1} = f$.

8. In Theorem 5.3.6 you proved that $f^{-1}(S^{\complement}) = (f^{-1}(S))^{\complement}$. Construct several examples that show that *absolutely nothing* can be said in general about the relationship between

$$f(S^{\complement}) \quad \text{and} \quad (f(S))^{\complement}.$$

9. Let $f : A \to B$ be a function. Suppose K and J are subsets of A. Let $U \subseteq B$.

(a) Prove that $f(K) \subseteq U$ if and only if $K \subseteq f^{-1}(U)$.

(b) Prove that if $K \subseteq J$, then $f(K) \subseteq f(J)$.

(c) Prove that it is not always the case that if $f(K) \subseteq f(J)$, then $K \subseteq J$.

(d) Complete and prove the following statement:

   If f is _____, *then $f(K) \subseteq f(J)$ implies that $K \subseteq J$.*

10. Let $f : A \to B$ be a one-to-one correspondence. Suppose that Ω is a partition of A. Prove that

$$\Omega^* = \{f(K) : K \in \Omega\}$$

is a partition of B.

11. Let $f : A \to B$ be a function. Let X and Y be subsets of A, and U and V be subsets of B.

(a) Prove that $f^{-1}(U) \setminus f^{-1}(V) = f^{-1}(U \setminus V)$.

(b) Prove that $f(X) \setminus f(Y) \subseteq f(X \setminus Y)$. Show that these two sets are equal for all subsets X and Y of A if and only if f is one-to-one.

12. Let $f : A \to B$ be a function. In this problem we will consider sets of the form $f(f^{-1}(S))$.

(a) Show that for all subsets S of B, $f(f^{-1}(S)) \subseteq S$.

(b) Give an example to show that $f(f^{-1}(S))$ need not be equal to S.

(c) Complete and prove the following statement:

   $f(f^{-1}(S)) = S$ *for all subsets S of B if and only if* _____.

13. Let $f : A \to B$ be a function. In this problem we will consider sets of the form $f^{-1}(f(T))$.

(a) Show that for all subsets T of A, $T \subseteq f^{-1}(f(T))$.

(b) Give an example to show that $f(f^{-1}(T))$ need not be equal to T.

(c) Complete and prove the following statement:

   $f^{-1}(f(T)) = T$ *for all subsets T of A if and only if* _____.

14. Let $f : A \to B$ be a function. For each $b \in B$, define $\mathcal{F}(b) = f^{-1}(\{b\})$. (Note that $\mathcal{F}$ is a function from B to $\mathcal{P}(A)$.)

(a) Show that $\mathcal{F}$ is one-to-one if f is onto.

(b) Give an example to show that if f is not onto, $\mathcal{F}$ need not be one-to-one.

(c) Can $\mathcal{F}$ ever be onto? Prove your conjecture.

15. Let $f : A \to B$ be a function. We can generate a function from $\mathcal{P}(A)$ to $\mathcal{P}(B)$ using images. Define $\mathcal{F} : \mathcal{P}(A) \to \mathcal{P}(B)$ by $\mathcal{F}(S) = f(S)$ for each $S \in \mathcal{P}(A)$. Under what circumstances is $\mathcal{F}$ one-to-one? onto? Prove your answer.

16. Let $f : A \to B$ be a function. We can generate a function from $\mathcal{P}(B)$ to $\mathcal{P}(A)$ using inverse images. Define $\mathcal{F} : \mathcal{P}(B) \to \mathcal{P}(B)$ by $\mathcal{F}(T) = f^{-1}(T)$ for each $T \in \mathcal{P}(B)$. Under what circumstances is $\mathcal{F}$ one-to-one? onto? Prove your answer.

17. Let A and B be two sets. Let $f : A \to B$ be a one-to-one correspondence. Show that $(\mathcal{P}(A), \subseteq)$ and $(\mathcal{P}(B), \subseteq)$ are order isomorphic. (*Hint:* Working Problems 9 and 15 first can help.)

18. Let $(A, \leq_A)$ and $(B, \leq_B)$ be order-isomorphic partially ordered sets. Let y and $x \in A$. Then y is the immediate successor (resp. immediate predecessor) of x in A if and only if $f(y)$ is the immediate successor (resp. immediate predecessor) of $f(x)$ in B.

19. Let

$$A = \left\{ x \in \mathbb{R} : x = 1 - \frac{1}{n} \right\}.$$

Let $B = A \cup \{1\}$, as in Exercise 5.4.10. Let $C = A \cup \{1, 2\}$. Prove that C is not order isomorphic to either A or B.

20. Let $(A, \leq_A)$ and $(B, \leq_B)$ be isomorphic partially ordered sets. Let J be a subset of B and let $x \in A$.

(a) x is a lower bound for $f^{-1}(J)$ in A if and only if $f(x)$ is a lower bound for J in B.

(b) x is the greatest lower bound for $f^{-1}(J)$ in A if and only if $f(x)$ is the greatest lower bound for J in B.

Parallel statements hold for upper bounds and least upper bounds. (You should formulate these statements and verify that your arguments for lower and greatest lower bounds can be adapted to prove them.)

21. Let $(A, \leq_A)$ and $(B, \leq_B)$ be isomorphic partially ordered sets. Prove that A has the least upper bound property if and only if B has the least upper bound property.

22. Consider the function $f : A \to A$ shown in Figure 5.8 (page 134). Find all iterated maps on f.

23. Iterated maps are very important in the theory of chaos. Of special interest is the **logistic map** $f : (0, 1) \to (0, 1)$, given by $f(x) = kx(1 - x)$ where $k \in (0, 4]$. The sequences obtained by iterating f have some fascinating properties. The most interesting fact is that the long-term behavior of these sequences depends mostly on the choice of k and very little on the choice of a_0! Try your hand at iterating f using various values of $a_0 \in (0, 1)$ for the following values of k. (Get a computer to do the arithmetic so you can easily generate lots of terms.)

(a) $k = 2.5$
   ▪ **Attractor of period 1:** Sequences converge.
   ▪ Evaluate the function f at $x = \frac{6}{10}$. What do you see?

(b) $k = 3.2$
   ▪ **Attractor of period 2:** Sequences eventually appear to oscillate between two values.
   ▪ Evaluate f at the points

$$x = \frac{21 + \sqrt{21}}{32} \approx 0.7994 \quad \text{and} \quad x = \frac{21 - \sqrt{21}}{32} \approx 0.5130.$$

(Compute using exact expressions, not decimal approximations.) What do you see?

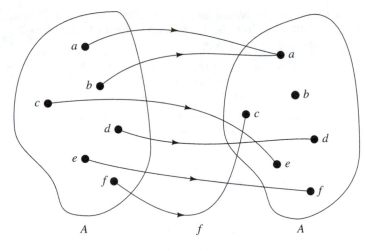

Figure 5.8

(c) $k = 3.5$ **Attractor of period 4:** Sequences eventually appear to oscillate among four values.

(d) $k = 4$ **Chaos:** Sequences should appear to fluctuate randomly.

(This problem is meant to be suggestive, not conclusive. See Question to Ponder 4.)

24. Let $*$ be binary operation on a set A. An element $e \in A$ is said to be an **identity** for $(A, *)$ provided that $e * a = a * e = a$ for all $a \in A$. Which of the binary operations defined in Section 5.6 have an identity? For those that do, identify it and prove that it is an identity.

25. Let X be a set. Recall the *symmetric difference* of two sets A and B:

$$A \triangle B = (A \setminus B) \cup (B \setminus A).$$

Show that $\triangle$ is a commutative and associative binary operation on $\mathcal{P}(X)$. Does $(\mathcal{P}(X), \triangle)$ have an identity (as defined in Problem 24)?

26. Just as a binary operation takes pairs of elements of A and gives in return an element of A, there are important operations, called **unary operations**, in which the input is an element of A and the output is also. For instance:

Of course, "unary operation on A" is just a fancy name for a function from A to A, but the name is used to underscore a particular point of view.

- Complementation on $\mathcal{P}(X)$ is a unary operation.
- The operation that takes a nonzero real number and returns its reciprocal is a unary operation on the nonzero real numbers.

A **Boolean algebra** $(X, *, \diamond, \prime)$ is a set X together with two binary operations $*$ and $\diamond$ and a unary operation $\prime$ on X satisfying the following properties for all $A, B, C \in X$:

(a) $*$ and $\diamond$ are commutative and associative.

(b) $*$ distributes over $\diamond$, and $\diamond$ distributes over $*$. That is:

$$A * (B \diamond C) = (A * B) \diamond (A * C) \text{ and } A \diamond (B * C) = (A \diamond B) * (A \diamond C).$$

(c) There is an identity 0 for $(X, *)$ called the *zero element*.

(d) There is an identity 1 for $(X, \diamond)$ called the *unit element*.

(e) $A \diamond A' = 0$ and $A * A' = 1$.

Prove that for all sets S, $(\mathcal{P}(S), \cup, \cap, ^{c})$ is a Boolean algebra.

27. A set G together with an associative binary operation $*$ is called a **group** if

   i. $(G, *)$ has an identity element e.

   ii. For all $a \in G$, there exists $a^{-1} \in G$ such that $a * a^{-1} = a^{-1} * a = e$.

   (a) Prove that $(\mathbb{N}, +)$ and $(\mathbb{R} \setminus \{0\}, \cdot)$ are groups.

   (b) Let X be any set. Prove that the set of all one-to-one correspondences from X to X is a group under function composition.

■ QUESTIONS TO PONDER

1. The following is a variation of a problem found in the book *Contemporary Precalculus* written by the North Carolina School for Science and Mathematics (Janson Publications). The precalculus students were asked to "play" with the function. Actually proving the answer requires complete induction. It's kind of tricky, but a lot of fun.
   Define $f : \mathbb{Z} \to \mathbb{R}$, as follows:

$$f(x) = \begin{cases} x - 10 & \text{if } x > 100, \\ f(f(x + 11)) & \text{if } x \le 100. \end{cases}$$

   Identify the range of f and prove your conjecture.

2. Ponder some more the questions on indexing and counting mentioned at the end of Chapter 2. Your knowledge of functions and sequences should help you to make additional progress on them. (In particular, a sequence of distinct terms is an example of an indexed set. Why?)

3. Following the train of thought arising from Question 2, suppose we have two *infinite* sets, A and B. What might it mean to say that A and B have the same number of elements? Is it possible to make any sense of such a notion? If so, do all infinite sets have the same number of elements?

4. Think about the very suggestive outcomes of Problem 23. Try iterating other functions from $\mathbb{R}$ to $\mathbb{R}$. Make and test conjectures. Try to prove them. *This is a great place to play. It is easy to make progress, but at the same time it can keep you busy indefinitely. There is a lot there!*

6 Elementary Number Theory

6.1 Natural Numbers and Integers

Number theory is devoted to the study of the arithmetic properties of the natural numbers and the integers. One might at first think that this is likely to consist primarily of trivialities that are apparent to any fifth grader, but this is definitely not so. Difficult and even unsolved problems abound in number theory. However, unlike the difficult problems in many other branches of mathematics, the problems of number theory can often be stated in such a way that anyone with a background in basic algebra can easily understand what is being asked. Furthermore, even problems that seem intractable to the greatest mathematicians can be played with by the amateur. You may remember the "Goldbach conjecture" that was mentioned in Chapter 1. Another famous unsolved problem is the "twin prime conjecture." Pairs of prime numbers that differ by two are called "twin" primes. For instance, 7 and 9, 29 and 31, 4517 and 4519, 16631 and 16633.

Twin Prime Conjecture: There are infinitely many pairs of twin primes.

Number theory is a subject in its own right, and we will hardly scratch the surface with what we do here. Our discussion will center on the theme of divisibility. In the introduction to the book, we said that *Chapter Zero* would serve as your road map on a journey. Well, this part of the journey will take a . . . scenic route. We will visit a number of ideas that are important for a rudimentary understanding of divisibility. After some discussion of divisibility in the integers, we use these ideas to build a new algebraic structure: $\mathbb{Z}_n$, the integers modulo n. The chapter will culminate in a discussion of divisibility in $\mathbb{Z}_n$.

As we have from the outset, we will take for granted the most elementary arithmetic and order properties of the natural numbers and the integers. When we first established these ground rules, we did not say exactly what we meant to assume. We have now covered enough ground to make it possible to be explicit about our assumptions, and since they will play a central role in this chapter, it is especially important to do so.

- $\mathbb{N}$ and $\mathbb{Z}$ are endowed with two commutative and associative binary operations called addition ($+$) and multiplication ($\cdot$).[1]

    Multiplication distributes over addition. That is, for all integers x, y, and z,

$$x(y + z) = (xy + xz).$$

- There are two "special" integers, 0 and 1. When 0 is added to any integer n, the result is n. When 1 is multiplied by any integer n, the result is, once again, n. That is, 0 is an identity for addition and 1 is an identity for multiplication. 0 and 1 are unique in this respect. That is, no other number exists that has either of these properties.

- $\mathbb{N}$ and $\mathbb{Z}$ are endowed with a total order $\leq$. Each element x in these sets has an immediate successor, $x + 1$. Each element x in $\mathbb{Z}$ has an immediate predecessor, $x - 1$, as does each element except 1 in $\mathbb{N}$.

- Addition and multiplication mesh with $\leq$ in the following familiar ways. For all integers x, y, and z,

$$x \leq y \text{ if and only if } x + z \leq y + z.$$

$$\text{If } z \in \mathbb{N}, \text{ then } x \leq y \text{ if and only if } x \cdot z \leq y \cdot z.$$

- And (as we said in Chapter 3) $\mathbb{N}$ satisfies the induction axiom.

Before we go any further, we need one more fundamental property of $\mathbb{N}$. We can use the Induction Axiom to prove that $\mathbb{N}$ is a *well-ordered set*.

6.1.1   DEFINITION

A totally ordered set in which every nonempty subset has a least element is said to be **well-ordered**.

6.1.2   EXERCISE

Show (by giving counterexamples) that under the usual order $\leq$, $\mathbb{Z}$ and the interval $[0, 1]$ are not well-ordered.   □

6.1.3   THEOREM (Well-ordering of $\mathbb{N}$)

$\mathbb{N}$ is well-ordered.

(*Hint:* Proceed by contraposition. Let K be a subset of $\mathbb{N}$ with no least element and show that K is empty by using induction to show that its complement is all of $\mathbb{N}$.)   □

6.1.4   THEOREM

Use the well-ordering of $\mathbb{N}$ to prove the following.

1. Every nonempty subset of $\mathbb{Z}$ that has a lower bound has a least element.

---

[1] As is customary, we will often omit the multiplication symbol and write ab instead of $a \cdot b$.

2. Every nonempty subset of $\mathbb{Z}$ that has an upper bound has a greatest element. □

Something to keep in mind: The well-ordering of $\mathbb{N}$ is especially relevant to us because it is essential to proving many important facts about the natural numbers and the integers. It is a standard proof technique to identify a nonempty subset S of those natural numbers that have a certain property. Because we are guaranteed that S has a least element ℓ, we can then use the existence of ℓ to achieve some purpose. This is useful in two general sorts of circumstances. The first is straightforward: In an existence proof the thing we are trying to pinpoint may be the smallest of all natural numbers (or integers) that have a particular property, as in the proof of the division algorithm (Theorem 6.2.1) in the next section. The second, less obvious, is in proof by contradiction. The trick is to use a supposed least element ℓ of a nonempty set $S \subseteq \mathbb{N}$ to construct another natural number $\ell' \in S$ that is even smaller than ℓ, thus forcing a contradiction. Keep these two approaches in mind as you try to prove the remaining theorems in this chapter.

Remark. The logical connection between the Induction Axiom and the well-ordering of $\mathbb{N}$ is very strong: Given our other assumptions about the existence of immediate successors and predecessors in $\mathbb{N}$, they are equivalent. Thus we could have taken the well-ordering of $\mathbb{N}$ as an axiom and shown that $\mathbb{N}$ also satisfies the Induction Axiom. (See Question 1 in the Questions to Ponder at the end of the chapter.)

6.2   Divisibility in the Integers

We can easily define and discuss the inverse operation of addition (subtraction) on $\mathbb{Z}$, but the same cannot be done with the inverse of multiplication, since when we divide one integer by another we do not always obtain an integral result. For instance, we can easily make sense of $\frac{8}{2}$ within the integers, but not $\frac{19}{7}$. So what *can* we say about divisibility in the integers?

The first step we take as we begin our discussion of divisibility is to establish our ability to divide any positive integer into any integer, obtaining a unique quotient and a unique remainder, both integers. It is not necessary to refer to division at all: One may interpret the statement $\frac{19}{7} = 2\frac{5}{7}$ to mean $19 = (2 \cdot 7) + 5$. This is what we mean when we say that 19 divided by 7 yields a quotient of 2 and a remainder of 5. These ideas are formalized in the *division algorithm*.

6.2.1   THEOREM (The division algorithm)

Let $m \in \mathbb{Z}$ and let $n \in \mathbb{N}$. Then there exist unique integers q and r such that

$$m = qn + r \quad \text{and} \quad 0 \le r < n.$$

The integer q is called the **quotient** and the integer r is called the **remainder**.

(*Hint:* You can proceed in one of two ways. Find a set whose largest element is q or one whose smallest element is r. Define the other in terms of the one you found. Once you have found q and r, don't forget to show that they are unique.)   □

6.2.2 EXERCISE

Apply the division algorithm to find quotients and remainders for the following pairs of numbers.

    1. $25 = $ _____ $3 + $ _____

    2. $36 = $ _____ $9 + $ _____

    3. $-10 = $ _____ $6 + $ _____ (Remember that r must be *positive*.)   □

6.2.3 EXERCISE

Let a and b be natural numbers. Suppose $a = qb + r$ with $0 \le r < b$. What does the division algorithm yield when $-a$ is divided by b? Justify your answer.   □

Those natural numbers n that when divided into m using the division algorithm yield a remainder of zero have a special status with respect to m. We say that n divides m evenly or simply that n divides m. This concept can be meaningfully defined for any two integers.

6.2.4 DEFINITION

Let m and n be integers. We say that m **is divisible by** n, that n **divides** m, or that n **is a divisor of** m if there exists an integer a so that $m = an$. We denote this by $n \mid m$.

6.2.5 EXERCISE

Let a, b, and $c \in \mathbb{Z}$. Prove or disprove: If $a \mid bc$, then $a \mid b$ or $a \mid c$.   □

The following properties of divisibility follow easily from the definition.

6.2.6 THEOREM

Let a, b, and $c \in \mathbb{Z}$. Then

    1. $a \mid a$.

    2. $1 \mid a$, and $-1 \mid a$.

    3. $a \mid 0$.

    4. If $b \mid a$, then $b \mid (-a)$.

    5. If $a \mid b$ and $b \neq 0$, then $|a| \le |b|$.   □

Some natural numbers have an especially simple divisibility structure. They play a central role in number theory.

6.2.7 DEFINITION

A natural number p is said to be a **prime number** if:

- $p > 1$, and
- the only positive divisors of p are 1 and p.

Natural numbers that are not prime are called **composite numbers**.

The next theorem shows that divisibility yields a natural partial order on $\mathbb{N}$.

6.2.8 THEOREM

Let a, b, and $c \in \mathbb{Z}$.

1. If $a \mid b$ and $b \mid a$, then $a = \pm b$.
2. If $a \mid b$ and $b \mid c$, then $a \mid c$. $\qquad\qquad$ □

6.2.9 COROLLARY

$\mathbb{N}$ is partially ordered under the relation $\mid$ ("divides"). (**Question:** Why $\mathbb{N}$ and not $\mathbb{Z}$?)
$\qquad\qquad$ □

The following problem is a reprise of Problem 10 at the end of Chapter 4. It will give you help in thinking about the partial ordering $\mid$. It will also motivate some subsequent thinking.

6.2.10 EXERCISE

Answer the following questions about the partial ordering $\mid$ on $\mathbb{N}$.

1. Prove that $\mathbb{N}$ is partially ordered under the relation $\mid$.
2. Is $\mid$ a *total* order on $\mathbb{N}$? Explain.
3. Draw a lattice diagram that depicts the order $\mid$ on the set $\{1, 2, 3, \ldots, 15\}$.
4. Does $\{2, 3, 4, 5, \ldots\}$ have any minimal or maximal elements (with respect to the order $\mid$)?
5. Complete the following sentence:

   An element $d \in \mathbb{N}$ is a lower bound for $S \subseteq \mathbb{N}$ under the order $\mid$ if and only if _____.

   Phrase a parallel statement that says what it means for d to be an upper bound for S.
6. Show that the set $\{12, 18\}$ has a greatest lower bound and a least upper bound in $(\mathbb{N}, \mid)$. What are they? (There are more common names for these. Do you know what they are?) $\qquad$ □

The third part of Exercise 6.2.10 suggests the following definitions.

6.2.11 DEFINITION

Let a, b, d, and $m \in \mathbb{Z}$. Then d is a **common divisor** of a and b if $d \mid a$ and $d \mid b$. m is a **common multiple** of a and b if $a \mid m$ and $b \mid m$.

Two integers are said to be **relatively prime** if their only positive common divisor is 1.

6.2.12 EXERCISE

1. Find a pair of numbers that has 7 as a common divisor.

2. Give an example of two composite numbers that are relatively prime.

3. Find two pairs of relatively prime numbers that have 54 as a common multiple.

$\square$

The following easy theorem is very useful and will help you work through the definitions.

6.2.13 THEOREM

Let a, b, c, d, x, and $y \in \mathbb{Z}$. If c is a common divisor of a and b, then $c \mid (ax + by)$. Similarly, if c and d are common multiples of a and b, then $cx + dy$ is a common multiple of a and b.

$\square$

6.2.14 EXERCISE

Assume that r and s are relatively prime. Show that $r + s$ and s are relatively prime, also.

$\square$

6.2.15 EXERCISE

Let a and $b \in \mathbb{Z}$.

1. If a and b are not both zero, show that there exists a unique largest (with respect to $\leq$) common divisor of a and b. Explain why this number must be positive.

2. This last statement is false if a and b are both zero. Explain why. (Be sure to notice where in your argument you used this fact.)

3. Show that there exists a unique smallest (with respect to $\leq$) *positive* common multiple of a and b.

$\square$

Exercise 6.2.15 allows us to speak of the numerically largest common divisor of a and b and the numerically smallest positive common multiple of a and b. However, the fact that these are numerically largest and smallest is not in itself sufficiently useful to merit much attention. The last part of Exercise 6.2.10 suggests something more interesting; they are greatest and smallest also under partial ordering "divides." (They are the greatest lower bound and the least upper bound of the set $\{a, b\}$ under $\mid$.)

6.2.16 THEOREM (The Least Common Multiple)

Let a and $b \in \mathbb{Z}$. Then the following statements are equivalent for a positive common multiple d of a and b.

    i. $d \leq c$ for all positive common multiples c of a and b.

    ii. $d \mid c$ for all common multiples c of a and b.

Since Exercise 6.2.15 showed that there exists a unique number satisfying i (and thus ii), we can now define d to be **the least common multiple of** a **and** b. We will denote it by $\mathrm{lcm}(a, b)$.

    (*Hint:* (i $\Rightarrow$ ii) Apply the division algorithm to divide c by d. Show that the remainder is also a common multiple of a and b. What does this tell you?) □

    We have a similar theorem for the greatest common divisor of a and b.

6.2.17 THEOREM (The greatest common divisor)

Let a and $b \in \mathbb{Z}$, not both zero. Then the following statements are equivalent for a common divisor d of a and b.

    i. $d \geq c$ for all common divisors c of a and b.

    ii. $c \mid d$ for all common divisors c of a and b.

    Since Exercise 6.2.15 showed that there exists a unique number satisfying i (and thus ii), we can now define d to be **the greatest common divisor of** a **and** b. We will denote it by $\gcd(a, b)$.

    (*Hint:* (i $\Rightarrow$ ii) Let c be any common divisor of a and b. What do the results of Theorem 6.2.16 tell you about the relationship between a and b and the least common multiple of c and d? What can you conclude?) □

6.2.18 EXERCISE

Let b and $c \in \mathbb{Z}$. Suppose that b and c are relatively prime. Show that for all integers a, $\gcd(a, b)$ and $\gcd(a, c)$ are relatively prime. □

6.2.19 LEMMA

Let a and b be natural numbers.

    1. You know that $\mathrm{lcm}(a, b) \mid ab$. (Why?) Show that if $ab = k\,\mathrm{lcm}(a, b)$, then k is a common divisor of a and b.

    2. You know that $\gcd(a, b) \mid ab$. (Why?) Show that if $ab = m\,\gcd(a, b)$, then m is a common multiple of a and b. □

    The following problem asks you to discover a relationship between the product of a pair of numbers, their greatest common divisor, and their least common multiple.

6.2.20   PROBLEM

1. Under what circumstances is the product of two natural numbers their least common multiple? Prove your answer.

2. Examine the entries in the following table.

a	b	ab	$\text{lcm}(a, b)$	$\gcd(a, b)$
2	6	12	6	2
−4	6	−24	12	2
6	8	48	24	2
6	−9	−54	18	3
24	36	864	72	12

Conjecture a relationship between the entries in the last three columns. Can you prove your conjecture? (*Hint:* Theorem 6.2.8(1) and Lemma 6.2.19 should be helpful.)   □

The relationship you found in Problem 6.2.20 tells us how to find either the greatest common divisor or the least common multiple if we are given the other. It doesn't tell us how to find either one to begin with. In the next section we will discuss a clever method for finding the greatest common divisor of a pair of numbers.

6.3   The Euclidean Algorithm

We now know that the greatest common divisor of any pair of integers exists and is unique. We are left with the problem of how to find it. If you were asked to find the greatest common divisor of 48 and 18, you could do it quickly by simple inspection. If you were asked for the greatest common divisor of 1,452,679 and 2,306,347, superficial inspection would not get you anywhere. (Trial and error could eventually succeed but would be tedious!) What we would like is to develop an algorithm, a well-specified repetitive procedure, that will eventually give us the greatest common divisor of any pair of numbers. Your inclination might be to try to factor the two numbers, but any algorithm that actually requires us to do that is not going to be useful for very large numbers. (Imagine trying to factor a number that is the product of two prime numbers each of which is six digits long!) The Euclidean algorithm allows us to calculate the greatest common divisor of two numbers fairly efficiently using only the division algorithm. It is based on the following simple theorem.

6.3.1   THEOREM

Let a and b be natural numbers. If

$$a = qb + r,$$

then the set of common divisors of a and b is the same as the set of common divisors of b and r. In particular, $\gcd(b, r) = \gcd(a, b)$. □

If $a > b$ and we obtain r by the division algorithm, b and r will be a new pair of numbers with the same set of divisors as the original pair. We have simplified our problem by turning it into an equivalent problem involving *smaller* numbers! Naturally, we need not stop there. We can repeat the procedure, dividing r into b to obtain a smaller number whose greatest common divisor with r is the same as the greatest common divisor of b and r (and thus of b and a). If we repeat this procedure over and over again, we obtain a sequence $q_1, q_2, q_3 \ldots$ of quotients and a sequence $r_1, r_2, r_3 \ldots$ of remainders satisfying:

$$a = q_1 b + r_1, \text{ where } 0 \le r_1 < b,$$
$$b = q_2 r_1 + r_2, \text{ where } 0 \le r_2 < r_1,$$
$$r_1 = q_3 r_2 + r_3, \text{ where } 0 \le r_3 < r_2,$$
$$r_2 = q_4 r_3 + r_4, \text{ where } 0 \le r_4 < r_3,$$
$$\vdots$$

In each case, the set of common divisors of r_i and r_{i+1} is the same as the set of common divisors of a and b. To see what happens, show the following.

6.3.2 EXERCISE

If we continue the procedure outlined above,

1. we must eventually obtain a remainder of zero, and
2. the last nonzero remainder obtained will be $\gcd(a, b)$. □

The algorithm outlined above for finding the greatest common divisor is called the **Euclidean algorithm**. Now you will have to review the discussion to see how to implement it. The best way to do this is just to follow the argument through some examples.

6.3.3 EXERCISE

Use the Euclidean algorithm to find the greatest common divisor for the following pairs of numbers.

1. 48 and 18.[2]
2. 2700 and 17,640
3. 1,452,679 and 2,306,347 □

---

[2] The point here is not to find the answer. (We all know it is 6.) The purpose is to try out the algorithm with a case in which you know the answer in advance. This is a good way to see whether you are implementing it correctly. Notice, however, that it is not a conclusive way. You could, in principle, get the right answer by chance even if you were doing something wrong; it's just not very likely.

6.3.4   QUESTION

Now you know how to calculate the greatest common divisor of a pair of natural numbers. How would you find the greatest common divisor of any pair of integers?

   (*Hint:* For natural numbers a and b, what is the relationship between $\gcd(a, b)$ and $\gcd(\pm a, \pm b)$?)   □

6.4   Relatively Prime Integers

You already know that if d is a common divisor of a and b, then $d \mid ax + by$ for all integers x and y (Proposition 6.2.13). If d is the greatest common divisor of a and b, more is true. It is possible to find integers x and y so that $d = ax + by$.

6.4.1   THEOREM

Let a and b be integers, not both zero. Let $d = \gcd(a, b)$. Then there exist integers x and y so that

$$d = ax + by.$$

(*Hint:* Let k be the least element of the set

$$S = \{am + bn : m, n \in \mathbb{Z} \text{ and } am + bn \geq 1\}.$$

Use proof by contradiction and the division algorithm to show that k is a common divisor of a and b. Then show that d is in S, also.   □

   This fact is surprisingly useful, especially when a and b are relatively prime, that is, when $\gcd(a, b) = 1$. (Every time you are allowed to assume that a pair of numbers is relatively prime, keep your eye out for a trick that uses this fact!)

6.4.2   COROLLARY

Let a and b be integers. The following are equivalent:

   1. a and b are relatively prime.
   2. There exist integers x and y such that

$$ax + by = 1.$$ □

   Relatively prime integers often behave much better with respect to products and divisibility than other pairs of integers. Here are some examples of such behavior.

6.4.3   THEOREM

Let a and b be relatively prime integers. Let c be any integer.

1. If $a \mid bc$, then $a \mid c$.

   (*Hint:* Write $c = c \cdot 1$.)

2. If $a \mid c$ and $b \mid c$, then $ab \mid c$. (Show by giving a counterexample that this may not be true if a and b are not relatively prime.) □

6.5 Prime Factorization

One of the most important theorems of arithmetic is the theorem that says that every natural number can be factored into a product of (one or more) primes and that this factorization will be unique up to the order of the factors. Because of its importance, this theorem is often called the fundamental theorem of arithmetic. The purpose of this section is to prove the fundamental theorem and to try to give you some sense of why it is so important.

You probably have some intuitive sense of why it is that small numbers can be written uniquely as products of primes. For numbers under 100 (or even 1000, if you are very diligent) you can probably check by hand to see that the fundamental theorem is true. But the theorem holds for all natural numbers, no matter how big. When very large numbers are involved, it is considerably less obvious. How do you *know* that 22,935,517,133 can be written as a product of primes? Well, you can probably convince yourself that either it is prime or it can be factored. If it can be factored, then its factors are either prime or can be factored, and if they can be factored, then their factors are either prime or *they* can be factored, and so on. Our goal, of course, is to get rid of the "and so on." We do this the way we have in the past—we use mathematical induction.

6.5.1 THEOREM

Every natural number is either prime or can be written as the product of primes. □

We have now established that natural numbers can be prime factored. But are those prime factorizations unique? Suppose that you and a friend were to go into separate rooms and work diligently until you each managed to prime factor 22,935,517,133. Do you think the two of you would come up with the same answer? Maybe there are 45 *different* ways to write 22,935,517,133 as a product of prime numbers. If this is not so, why is it not so—and why is this significant, anyway?

If you wanted to decide whether 1400 and 3267 were relatively prime, how would you proceed? Well, one of the easiest ways would be to prime factor the two numbers: $1400 = 2^3 5^2 7$ and $3267 = 3^3 11^2$. We can now observe that these numbers are relatively prime since they do not share a prime divisor . . . or can we? Answer: Not unless we can be sure that these numbers do not have other prime factorizations! If we are not guaranteed that prime factorizations are unique, the fact that $1400 = 2^3 5^2 7$ would *not* tell us that 1400 is not divisible by 11, nor would these factorizations tell us that 17 is not a common divisor of 1400 and 3267. It is the uniqueness of prime factorizations that drives our ability to draw conclusions from them! (Never underestimate the importance of a uniqueness theorem!)

The uniqueness clause in the fundamental theorem is, in fact, the heart of the theorem. If prime factorizations were not unique, their existence would itself be of little interest. Furthermore, this part of the theorem is considerably less obvious than the other. How do we know that you and your friend, closeted off by yourselves, will come up with the *same* factorization? It is all driven by the following little lemma.

6.5.2   LEMMA

Let p be a prime number.

1. Let m and n be integers. If $p \mid mn$, then $p \mid m$ or $p \mid n$.

From this it follows that

2. if $m_1, m_2, m_3, \ldots, m_k$ are integers, and if $p \mid m_1 m_2 m_3 \ldots m_n$, then $p \mid m_i$ for some i.   □

6.5.3   THEOREM   (The fundamental theorem of arithmetic)

Every positive integer can be written as a product of prime numbers, and that factorization is unique up to the order of the factors.

(*Hint:* Prove by induction that for any integer that can be written as a product of k primes, factorization is unique up to the order of the factors. This procedure will count the same prime factor as many times as it shows up in the factorization. For instance, if we write $1400 = 2^3 5^2 7$, we have written it as a product of six primes. This is called **counting according to multiplicity**.)   □

Many of the theorems that you have proved about greatest common divisors and so forth can alternatively be proved using the fundamental theorem of arithmetic. Certainly, much of our intuition about them and other facts about divisibility comes from it.

6.5.4   EXERCISE

Define the greatest common divisor and the least common multiple in terms of the prime factorizations of two numbers.   □

6.6   Congruence Modulo n

In the section on equivalence relations (Exercise 4.3.23, part 3) you showed that the relation $a \sim b$ if and only if $a - b$ is a multiple of 5 is an equivalence relation on the set of integers. Your proof did not depend on any property of the number 5. Indeed, this phenomenon is quite general.

6.6.1 EXERCISE

Show that if $n \in \mathbb{N}$, then the relation "$a \sim b$ if and only if $a - b$ is divisible by n" is an equivalence relation on $\mathbb{Z}$. □

6.6.2 DEFINITION

We will say that two integers are **congruent (or equivalent) modulo** n provided that their difference is divisible by n. This equivalence relation is called **congruence modulo** n. The equivalence classes of congruence modulo n are called the **congruence classes modulo** n. If a is equivalent to b modulo n, we denote this by $a \equiv b \pmod{n}$.

6.6.3 EXERCISE

a. Find three integers x, y, and z such that $x = 7 \pmod 6$, $y = 7 \pmod 6$, and $z = 7 \pmod 6$.

b. Describe the congruence classes modulo 6 as completely as possible. □

Our use of the term "equivalent" when talking about two integers whose difference is divisible by n is naturally evoked by the fact that congruence modulo n is an equivalence relation. But Exercise 6.6.3 suggests that we can give a more natural interpretation to this equivalence. For each natural number n, the division algorithm partitions the integers into n subsets: those integers that are divisible by n, those that yield a remainder of 1 when divided by n, those that yield a remainder of 2, a remainder of 3, . . . , a remainder of $n - 1$. Suppose we declare two integers to be equivalent if the division algorithm yields the same remainder when they are divided by n. Then we get an equivalence relation. This equivalence relation is, in fact, congruence modulo n, and the congruence classes modulo n are just the collections of integers that yield the same remainder when divided by n.

6.6.4 EXERCISE

Let $a, b \in \mathbb{Z}$. Let $n \in \mathbb{N}$. Show that the following two statements are equivalent.

1. $n \mid (a - b)$

2. The division algorithm yields the same remainder when a and b are divided by n. □

Thus we can determine whether $a \equiv b \pmod{n}$ either by seeing whether n divides their difference or by applying the division algorithm to them both and comparing remainders. The two concepts are the same.

The set of congruence classes modulo n is called the **integers modulo** n and is denoted by $\mathbb{Z}_n$. For each integer a we will denote the congruence class that contains a by $\bar{a}$.

Because there are many elements in each congruence class, each class will have many names. For instance, if we consider the congruence classes mod 6, $\overline{-18}$, $\overline{12}$, and $\overline{36}$ all denote the *same congruence class*! Similarly, $\overline{4} = \overline{52} = \overline{-2}$, and so forth. Be careful not to assume elements are different simply because they are referred to in different ways.

Because each congruence class contains the remainder with which it is identified,

$$\mathbb{Z}_n = \{\overline{0}, \overline{1}, \overline{2}, \ldots, \overline{n-1}\},$$

and therefore we call the set $\{0, 1, 2, \ldots, n-1\}$ a **complete set of congruence class representatives.**

In general, any set containing a single representative from each class is called a **complete set of congruence class representatives.** (Not an inventive name, but it does the job!) Of course, there are many such sets. Here are a few for $\mathbb{Z}_6$.

$$\{-18, -5, -10, 15, 4, 23\},$$
$$\{0, 7, -16, 21, -2, 17\},$$
$$\{12, -17, 8, 3, -14, -19\}.$$

However, none of these sets is very convenient. The set of possible remainders—$\{0, 1, 2, \ldots, n-1\}$—is much easier to remember and easier to deal with in computations, as you will see. Hence we will ordinarily use this complete set of representatives when one is called for. One might call it the **principal set of congruence class representatives.**

There are many partitions of $\mathbb{Z}$, of course, but from an arithmetic and algebraic point of view, $\mathbb{Z}_n$ is very special. It allows us to define addition and multiplication on the set of congruence classes in a natural way. The operations are performed by picking a representative from each class and allowing the sum or product of the pair of elements of the classes to represent the sum or product of the pair of classes. Here is the procedure:

- Take two congruence classes mod n.
- Take an element from each of them.
- Add (or multiply) these two elements together in the usual way.
- Find the congruence class that contains the result.
- Define this congruence class to be the sum (or product) of the two original congruence classes.

Now, you might think that our ability to implement this procedure did not depend at all on the fact that the partition being used was congruence modulo n. The same procedure could be used for any partition of $\mathbb{Z}$. After all, as long as the resulting sum (or product) of the chosen pair of numbers is in some class (and it is, since the sets in the partition exhaust the integers) and as long as it is in only one class (and it is, since distinct elements of the partition don't intersect), we should be able to perform the procedure.

This is all true, but there is an additional factor that must concern us. It is very important that a binary operation be **well-defined**—that the operation be sufficiently well described so that for any pair of objects, the operation will yield a *single, unambiguous result.* That is, it must be the case that if Joe comes along today and adds two objects and if Karen comes along tomorrow and adds the same two objects, Joe and Karen will

> Remember, a binary operation is a function, so for each input there must be one and only one output.

obtain the same result. It is not at all obvious that our procedure will guarantee this outcome. Our description allows Joe and Karen to pick any pair of elements, one from each of the classes. If they happen to pick different pairs (which they are likely to do), who knows if they will get the same result? The sums of the two pairs that they pick must be in the same congruence class. If they are not, addition will not be well-defined. (Likewise for multiplication.) For most partitions of $\mathbb{Z}$, addition and multiplication of classes (as defined above) are not well-defined.

6.6.5   EXERCISE

To get a better feel for all this, try the following exercise.

1. Pick any two congruence classes mod 6 from the list that you made in Problem 6.6.3. Pick three elements from each class. Add the pairs of elements together and verify that the results all end up in the same congruence class mod 6. Do the same for multiplication.

2. Partition $\mathbb{Z}$ into the three sets:

$$A = \{z \in \mathbb{Z} : z \le -10\},$$
$$B = \{z \in \mathbb{Z} : -9 \le z \le 9\},$$
$$C = \{z \in \mathbb{Z} : z \ge 10\}.$$

   Show that the operations of addition and multiplication (as we described them above) are not well-defined on this partition. (*Hint:* Just try taking pairs of elements and adding or multiplying them.)   □

The following theorem shows that addition and multiplication modulo n are well-defined on the congruence classes modulo n. (Be sure you understand why the well-definedness of $+_n$ and $\cdot_n$ follows from the theorem.)

6.6.6   THEOREM

Let a, b, c, and d be in $\mathbb{Z}$. Suppose that $a \equiv c \pmod{n}$ and $b \equiv d \pmod{n}$. Then:

    1. $a + b \equiv c + d \pmod{n}$, and

    2. $a \cdot b \equiv c \cdot d \pmod{n}$.   □

Now that we know the operations make sense, we can define them formally.

6.6.7   DEFINITION   (Addition and multiplication mod *n*)

We define the binary operations **addition modulo n** ($+_n$) and **multiplication modulo n** ($\cdot_n$) on the congruence classes mod n as follows. Let $\overline{a}$ and $\overline{b}$ be two congruence classes mod n. Then we define

> We place a subscript n on our new operations of addition modulo n and multiplication modulo n to distinguish them from ordinary addition and multiplication of integers.

$$\overline{a} +_n \overline{b} = \overline{a + b} \quad \text{and} \quad \overline{a} \cdot_n \overline{b} = \overline{a \cdot b}.$$

We are now in a position to see more clearly the use of the complete set of congruence class representatives that we described on page 150. Suppose, for instance, that we want to calculate $\overline{5}^{100}$ in $\mathbb{Z}_6$. We could, of course, calculate 5^{100} and then divide by 6 to find the remainder, but it would be easier to revert back to one of the primary congruence class representatives after each multiplication.

6.6.8   EXAMPLE

$$\overline{5}^2 = \overline{5} \cdot_n \overline{5} = \overline{25} = \overline{1}$$

$$\overline{5}^3 = \overline{1} \cdot_n \overline{5} = \overline{5}$$

$$\overline{5}^4 = \overline{5} \cdot_n \overline{5} = \overline{25} = \overline{1}$$

$$\vdots \quad\quad \vdots \quad\quad \vdots \quad\quad \vdots$$

Notice the interesting repeating pattern. Complete the example. Find $\overline{5}^{100}$ in $\mathbb{Z}_6$.   ■

> The fact that the pattern repeats is not a coincidence. If you raise any congruence class mod n to successive powers, a repeating pattern will eventually establish itself. Can you see why? Can you prove it?

6.7   Divisibility Modulo *n*

So how much of the structure of arithmetic carries over into $\mathbb{Z}_n$ with $+_n$ and $\cdot_n$? Many of the arithmetic properties that you are familiar with carry over here and often follow trivially from their counterparts in integer arithmetic. Here are a few of the most important.

6.7.1   THEOREM

    1. $+_n$ and $\cdot_n$ are commutative and associative operations.

2. $\cdot_n$ distributes over $+_n$.

3. $\overline{0} +_n \overline{a} = \overline{a}$ for all $\overline{a} \in \mathbb{Z}_n$.

4. $\overline{1} \cdot_n \overline{a} = \overline{a}$ for all $\overline{a} \in \mathbb{Z}_n$.

5. For all $\overline{a} \in \mathbb{Z}_n$, there exists $\overline{b} \in \mathbb{Z}_n$ such that $\overline{a} +_n \overline{b} = \overline{0}$. (Of course, $\overline{-a}$ is one example, but if you were to confine yourself to the primary set of congruence class representatives mod n, what would $\overline{b}$ be?)

6. For all $\overline{a}$, $\overline{b}$, and $\overline{c} \in \mathbb{Z}_n$, if $\overline{a} +_n \overline{b} = \overline{c} +_n \overline{b}$, then $\overline{a} = \overline{c}$. $\square$

However, there are some arithmetic properties that do not carry over except in special circumstances.

6.7.2   EXERCISE

Let a, b, and $c \in \mathbb{Z}_n$ with $\overline{b} \neq \overline{0}$. Show by giving a counterexample that it is not in general true that

$$\overline{a} \cdot_n \overline{b} = \overline{c} \cdot_n \overline{b} \qquad \text{implies} \qquad \overline{a} = \overline{c}.$$

In other words, it is not always possible to "cancel" a common factor from both sides of a multiplicative equation in $\mathbb{Z}_n$. $\square$

We will have more to say about this seeming anomaly later. For now, let us turn to the problem of divisibility. Divisibility in $\mathbb{Z}_n$ is, in some ways, less well behaved than in $\mathbb{Z}$, but it has an amazingly nice resolution in a special case.

Suppose that we are trying to make sense out of the fraction $\frac{10}{5}$ in $\mathbb{Z}_{15}$. Our first thought would be to guess that this is $\overline{2}$ when, in fact, this fraction is not well-defined. To see this, note that $10 \equiv 40 \pmod{15}$. Now, $\frac{10}{5} = 2$ and $\frac{40}{5} = 8$, yet 2 and 8 are not in the same congruence class (mod 15), so different choices from the congruence class $\overline{10}$ yield different answers! That is, even divisions that make perfect sense in $\mathbb{Z}$, cannot necessarily be formed in $\mathbb{Z}_n$.

Let us back up a little and think about what it means to make sense of the fraction $\overline{a}/\overline{b}$. If we can find a *unique* solution $\overline{x}$ to the equation

$$\overline{a} = \overline{b} \cdot_n \overline{x},$$

we can simply define $\overline{a}/\overline{b}$ to be that solution. (If there is more than one solution to the equation above, $\overline{a}/\overline{b}$ will not be well-defined; we will be unable to choose between the various solutions.) Conversely, any sensible definition for $\overline{a}/\overline{b}$, will certainly satisfy this equation. Thus the ability to uniquely solve linear equations is at the heart of the divisibility question. Unfortunately, solutions to

$$\overline{a} = \overline{b} \cdot_n \overline{x}$$

in $\mathbb{Z}_n$ may or may not exist, and when they exist, they may or may not be unique.

6.7.3 EXERCISE

Consider, for instance, the equation $\overline{6} = \overline{4} \cdot_n \overline{x}$. Show that it:

1. has no solutions in $\mathbb{Z}_8$.
2. has two solutions in $\mathbb{Z}_{10}$.
3. has a unique solution in $\mathbb{Z}_{15}$. □

Since in general we cannot expect solutions, unique or otherwise, to linear equations, our only hope is to look for restricted hypotheses that will guarantee the existence of unique solutions.

Whenever there *is* a solution to the equation, our ability to show it is unique boils down to our ability to say:

$$\overline{b \cdot x_1} = \overline{b \cdot x_2} \quad \text{implies} \quad \overline{x_1} = \overline{x_2}.$$

Because the "cancellation property" holds in $\mathbb{Z}$, linear equations that have solutions always have unique solutions. That is why the question of well-definedness for division never came up in our discussion of divisibility in $\mathbb{Z}$.

> Remember, if we know a solution exists and we wish to show it is unique, we assume there are two solutions and show they must be equal.

What about our ability to find a solution in the first place? We would clearly be home free if we could find an element $\overline{c}$ in $\mathbb{Z}_n$ so that $\overline{c} \cdot_n \overline{b} = \overline{1}$. (*Why?*) That is, we need to have a **multiplicative inverse** for b in $\mathbb{Z}_n$. Exercise 6.7.4 and Theorem 6.7.5 tell us under what circumstances we can do this.

> If $\overline{b}$ and $\overline{c} \in \mathbb{Z}_n$ and $\overline{b} \cdot_n \overline{c} = \overline{1}$, then b and c are said to be **multiplicative inverses** of one another.

6.7.4 EXERCISE

Which elements in $\mathbb{Z}_{10}$ have multiplicative inverses? Which elements in $\mathbb{Z}_{12}$ have multiplicative inverses? Which elements in $\mathbb{Z}_7$ have multiplicative inverses? □

6.7.5 THEOREM

Let $n \in \mathbb{N}$. Let $b \in \mathbb{Z}$. Then there exists $c \in \mathbb{Z}$ satisfying $\overline{c} \cdot_n \overline{b} = \overline{1}$ if and only if b and n are relatively prime.

(*Hint:* Don't forget Corollary 6.4.2.) □

Thus whenever b and n are relatively prime, we are guaranteed a solution to the equation

$$\overline{a} = \overline{b} \cdot_n \overline{x}$$

for all integers a. In fact, that solution will be unique.

6.7.6 THEOREM

Let $n \in \mathbb{N}$. Let $b \in \mathbb{Z}$. If there exists $c \in \mathbb{Z}$ satisfying $\overline{c} \cdot_n \overline{b} = \overline{1}$, then the equation

$$\overline{a} = \overline{b} \cdot_n \overline{x}$$

has a unique solution.   □

6.7.7 COROLLARY

If p is a prime number and $b \not\equiv 0 \bmod p$, then

$$\overline{a} = \overline{b} \cdot_p \overline{x}$$

has a unique solution for all integers a.   □

We have just gathered a fairly amazing fact: If b and n are relatively prime, then every element of $\mathbb{Z}_n$ is evenly divisible by b. In fact, if n is prime, then any congruence class modulo n is divisible by any other congruence class modulo n except $\overline{0}$. Division works beautifully in $\mathbb{Z}_p$, just as it does in the rational numbers or the real numbers!

■ **PROBLEMS**

1. Prove that the sum of the cubes of any three consecutive natural numbers is divisible by 9. (*Hint:* Use induction.)

2. I said in Section 6.5 that many theorems about divisibility are made easier by the fundamental theorem of arithmetic.

   (a) Find two theorems that you proved before the Section 6.5 whose proof you think is made easier by the fundamental theorem.

   (b) In the context of the fundamental theorem, think about divisibility. Formulate and prove a theorem that did not appear in this chapter.

3. Among partitions of $\mathbb{Z}$, $\mathbb{Z}_n$ is rather special. The following theorem says just how special it is:

   Let Ω be any partition of $\mathbb{Z}$. Define $+_\Omega$ on Ω by choosing representatives, adding them, and obtaining as a result the element of Ω that contains their sum. If $+_\Omega$ is well-defined, then either Ω is the trivial partition (each element of Ω contains a single element and Ω is just $\mathbb{Z}$ in disguise) or Ω is the partition given by equivalence modulo n for some $n \in \mathbb{N}$.

   In effect, if the addition of equivalence classes is well-defined, and the partition is nontrivial, then the equivalence relation *must* be congruence modulo n for some n. However, the same is not true for multiplication. As you will see in the following problem (though they are rare), there *are* nontrivial equivalence relations on $\mathbb{Z}$ other than congruence modulo n that yield a well-defined multiplication on equivalence classes.

4. Prove by giving an example that there are partitions Ω of $\mathbb{Z}$ other than $\mathbb{Z}_n$ and the trivial partition that yield a well-defined multiplication $\cdot_\Omega$. (*Hint:* One possible tack is to think about prime numbers, composites, and prime factorizations. What will multiplication do?)

■ QUESTIONS TO PONDER

1. Show that the well-ordering of $\mathbb{N}$ implies the Induction Axiom. (*Hint:* Refer back to the Induction Axiom on page 58. Your task is to show that any subset S of $\mathbb{N}$ satisfying the two provisions of the hypothesis must be all of $\mathbb{N}$. Proceed by contradiction—if S is not all of $\mathbb{N}$, there must be some smallest member of $\mathbb{N}$ that is not in S.

   The tricky part here is deciding what is an assumption and what is to be proved. The *logical structure* of the Induction Axiom (an implication within an implication) may throw you off the track. Remember: Everything before "then $S = \mathbb{N}$" is your hypothesis.)

2. Corollary 6.4.2 says that the converse of Theorem 6.4.1 is true for any pair of relatively prime integers. Is it true for any pair of integers? Why or why not?

3. In Exercise 6.2.5 you saw that, in general, if a divides bc, it need not divide either b or c. But in Lemma 6.5.2 you showed that if a is prime, then a must divide either b or c. In fact, prime numbers can be characterized in this way. A number that always "splits" products in this way can be shown to be prime. Formulate this principle precisely as a statement and then try your hand at proving it.

7 Cardinality

7.1 Galileo's Paradox

With wonder and consternation, philosophers and mathematicians have for centuries pondered the concept of infinity. Strange things happen in the realm of the infinite; much of our intuition simply breaks down there. One of the more famous of the paradoxes arising from the study of infinite sets was described by Galileo Galilei in the seventeenth century. In order to understand it, we must pause for a moment and consider how we decide whether two finite sets have the same number of elements.

Suppose we have a bin of nuts and a bin of bolts, and we want to know whether we have the same number of nuts and bolts. So we take an empty bin and begin drawing out one nut and one bolt at a time, screwing each pair together and putting the assemblies in the bin. If we run out of nuts and bolts at the same time, we can conclude that the numbers of nuts and bolts are the same, otherwise there are more of one than the other. If you think mathematically about what we have done here, we have attempted to construct a one-to-one correspondence between the set of nuts and the set of bolts. The pairs that we take are, in fact, the pairs of the function. If the nuts and bolts run out at the same time, then we get a one-to-one correspondence.

The language is more modern than that used by Galileo, but the idea is the same. He thought it reasonable to say that two sets are the same size if there is a one-to-one correspondence between them. Because we can build functions between sets of any size, the description need not be confined to finite sets. Galileo decided to apply it to infinite sets.

Galileo showed that there is a simple one-to-one correspondence between the set of natural numbers and the set of perfect squares (see Figure 7.1). On the other hand, he objected, there are clearly many more natural numbers than there are perfect squares; no set can have the same number of elements as one of its proper subsets. Thinking the

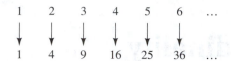

Figure 7.1   Galileo's paradox

situation unacceptably bizarre and contradictory, Galileo came to the conclusion that the labels "larger than," "smaller than," and "equal in size to" simply have no meaning when applied to infinite sets.

Nearly three centuries later, Georg Cantor, looking at the very same ideas, had a different perspective. He thought that Galileo's first insight had been correct. Cantor argued that it *is* reasonable to say that two sets are of the same size if there is a one-to-one correspondence between them, and that, as this is so, we must accept the fact that the set of natural numbers and the set of perfect squares are, indeed, sets of the same size. The fact that the definition behaves differently for infinite sets is not a flaw in the definition, but an insight into the nature of infinite sets.

However, if $\mathbb{N}$ is of the same "size" as one of its proper subsets, it is clear that notions of size are somewhat different for infinite sets than they are for finite sets; hence we do not use the word "size." We use instead the technical term *cardinality*.

7.1.1   DEFINITION

Let A and B be sets. A and B are said to have the same **cardinality** if there is a one-to-one correspondence between them. When A and B have the same cardinality, we denote this by card $A =$ card B.

7.1.2   EXERCISE

1. Show (by building the appropriate one-to-one correspondences) that the natural numbers, the even natural numbers, and the integers all have the same cardinality.

2. Let a, b, c, and d be real numbers with $a < b$ and $c < d$. Show that $[a, b]$ has the same cardinality as $[c, d]$. Likewise, (a, b) has the same cardinality as (c, d). What about (a, b) and $[c, d]$; do you think they have the same cardinality?   □

Despite Galileo's conclusions to the contrary, there are many ways in which our notion of what it means for two sets to have the same cardinality behaves very much like our notion of what it means for two finite sets to have the same number of elements.

7.1.3   THEOREM

Let A, B, and C be sets.

1. A has the same cardinality as itself.

2. If A has the same cardinality as B, then B has the same cardinality as A.

3. If A has the same cardinality as B and B has the same cardinality as C, then A has the same cardinality as C.

From these three facts we can conclude that if X is any set, then "has the same cardinality as" is an equivalence relation on $\mathcal{P}(X)$.   □

In order to extend our notions about cardinality, let us revisit the nuts and bolts analogy. If we run out of bolts before we run out of nuts, we see that there are more nuts than bolts. The function constructed in this case is a one-to-one function from the set of bolts into the set of nuts.

7.1.4   DEFINITION

If there is a one-to-one function from A into B, then we say that the cardinality of A is less than or equal to the cardinality of B and we denote this by card $A \leq$ card B.

Do the statements in the following theorem conform to your intuition about the size of sets?

7.1.5   THEOREM

Let A, B, and C be sets.

1. If $A \subseteq B$, then card $A \leq$ card B.
2. If card $A \leq$ card B, and card $B \leq$ card C, then card $A \leq$ card C.
3. If $C \subseteq A$ and B has the same cardinality as C, then card $B \leq$ card A.   □

> Suppose that for a finite set A, we denote the number of elements of A by $\#A$. Try this experiment in Theorems 7.1.3 and 7.1.5. Substitute
>
> - *finite set* for *set*.
> - $\#A = \#B$ for card $A =$ card B.
> - $\#A \leq \#B$ for card $A \leq$ card B.
>
> What does your intuition about finite sets tell you?
> Using this trick throughout the chapter will develop your intuition about cardinality and help you see where the notions of "cardinality" (for infinite sets) and "number of elements" (for finite sets) coincide and where they differ.

It is dangerous to carry the counting analogy too far. Following our intuition about comparing the number of nuts and bolts, we should be tempted to say that if the nuts run out before the bolts, then there are *more* bolts than nuts—in terms of functions, this would say that if there is a one-to-one function from A into B that is not onto, then B has *more* elements than A. While this is true for finite sets, Definition 7.1.4 stopped short of making an analogous statement for infinite sets. This is why.

7.1.6   EXERCISE

Give examples of sets A and B that have the same cardinality despite the fact that there is a one-to-one function from A into B that is not onto.   □

We have been using notation that is very reminiscent of the notation used to compare real numbers:

$$\text{card } A = \text{card } B \quad \text{and} \quad \text{card } A \le \text{card } B.$$

The choice of notation, together with the insight gathered in Exercise 7.1.6, make the following definition almost irresistible.

7.1.7 DEFINITION

Let A and B be sets. Then we say that card $A <$ card B if

- card $A \le$ card B, and
- it is not true that card $A =$ card B. (We might write this as card $A \ne$ card B.)

7.1.8 EXERCISE

Despite our use of the notation, we do not have an ordering; card $A \le$ card B and card $A =$ card B are symbolic ways of asserting the existence of certain functions from A to B. Thus the notion given in Definition 7.1.7 is not just the traditional interpretation of the symbol $<$ in terms of $\le$ and $\ne$. It is a statement about functions from A to B.

1. Recast Definition 7.1.7 explicitly in the language of functions.

2. Carefully explain the difference between Definition 7.1.7 and the statement

   $W :=$ "*There exists a function $f : A \to B$ that is one-to-one but not onto.*"

   In particular, be sure your explanation clearly shows that the condition given in Definition 7.1.7 is *stronger* than W. □

Despite my reminder that card $A \le$ card B is just a symbolic way of asserting the existence of certain functions between A and B, you probably suspect that the use of the notation is not a coincidence and you may be asking yourself whether there is an ordering lurking somewhere. As it turns out, there is, but making the notion precise is a bit tricky, and we will delay our discussion of this until later in the chapter. Meanwhile, I leave you with some questions to mull over as you continue your study of infinite sets and cardinality.

7.1.9 QUESTION

1. If we think of a partial ordering as a set of "comparisons," what exactly would we be comparing?[1] (If you think of finite sets, what is being compared when we say $\#A \le \#B$?)

2. Given an answer to our first question, what sorts of things would we need to do in order to prove that $\le$ is a partial ordering? What about a total order? Try to phrase these requirements in the language of functions. □

---

[1] This is the really tricky question!

7.2 Infinite Sets

Our purpose in this chapter is to explore the concept of cardinality and see where it takes us. But we cannot proceed with that yet, for we have not fully dispensed with Galileo's paradox. Galileo was no fool. We have to do more to dispel his doubts (and yours?) than just dismiss them out of hand. To see that Cantor's view should prevail, we must first examine what we mean when we say that a set is infinite.

7.2.1 DEFINITION

A set A is said to be **finite** if $A = \emptyset$ or if there is some $n \in \mathbb{N}$ so that A can be put into one-to-one correspondence with $\{1, 2, \ldots, n\}$.

7.2.2 DEFINITION

A set A is **infinite** if it is not finite.

To test this definition and because we will build on the proof later, we prove the following theorem. (If our definition is worth anything, this theorem certainly *ought* to be true!)

7.2.3 THEOREM

$\mathbb{N}$ is an infinite set.

Hint for proving Theorem 7.2.3: Proceed by contradiction. Assume that $f : \mathbb{N} \to \{1, 2, 3, \ldots, k\}$ is a one-to-one correspondence. Consider $f^{-1}(\{1\})$, $f^{-1}(\{2\})$, ..., $f^{-1}(\{k\})$. Each of these sets contains a single natural number. (Why?) Call these numbers $n_1, n_2, \ldots, n_k$, respectively. Let $n = n_1 + n_2 + \cdots + n_k$. Show that $n \notin \mathcal{D}om(f)$.

Actually, this argument doesn't work if $k = 1$. (Why not?) You will have to handle that case separately. □

Definition 7.2.2 tells us that if we get rid of all finite sets, we are left with the infinite ones. It tells us what infinite sets are not; it does not tell us what they are. We have said before that mathematical statements are almost always more useful if they are phrased positively rather than negatively. In preparation for reinterpreting Definition 7.2.2 in positive terms, we will examine some intuitive ideas about infinite sets by imagining a pair of scenarios.

FIRST SCENARIO: Picture a large basket full of things. Suppose we pull these things out one by one and discard them. If we eventually run out of things, we say that the set of things in the basket is finite; if we never run out, there are infinitely many things in the basket. Suppose we "count" the things as we take them out of the basket, assigning the label "one" to the first thing we remove, "two" to the second thing, and so forth. If the set is infinite and we never run out of elements, then we have a first element, a second element, a third element, and so on *ad infinitum*. That is, we have a *sequence*. Since we discarded each element after labeling it, the sequence must have distinct terms. We might thus assert that a set is infinite if it contains a sequence of distinct terms.

SECOND SCENARIO: Here is a variation of a tale told by the famous mathematician David Hilbert (1862–1943). Hilbert described a hotel with infinitely many rooms. It has a first room, a second room, a third room, and so on. On one stormy night when Hilbert's Hotel is full, a very wet and miserable couple come in and ask for a room. The concierge sees their plight and thinks for a minute. "All our rooms are full, but I will tell you what I can do. I can move the guests from room 1 into room 2, the guests from room 2 into room 3, and (in general) for each $n \in \mathbb{N}$, move the guests from room n into room $n + 1$. Then room 1 will be vacant for you."

7.2.4   EXERCISE

Later, a sequence of distinct visitors lines up at the reception desk. Hilbert's Hotel has a firm policy that no guest should have to change rooms more than twice in the middle of the night. What should the concierge do?

(*Hint:* You might think about what Galileo would do if he were the concierge.)   □

You have probably already noticed that the tale of Hilbert's Hotel is just a modern rendition of Galileo's paradox. However, it gives us some intuition about what was going on in Galileo's example. As it turns out, because of our ability to "shift things around," infinite sets *always* have the property that they can be put into one-to-one correspondence with a proper subset of themselves. Cantor was right. The example that made Galileo reject his original insight is the very example that reveals the nature of infinity.

7.2.5   THEOREM

Let A be a set. The following statements about A are equivalent.

i. A is infinite.

ii. A contains a sequence of distinct terms.

iii. A can be put into one-to-one correspondence with a proper subset of itself.   □

Hints for Proving Theorem 7.2.5: This theorem makes rigorous our intuitive notions about infinite sets. You must be careful to avoid using those intuitive ideas to justify your reasoning. The following hints should help you to do that.

First of all, it is probably easiest to prove i ⟺ ii and ii ⟺ iii. It is difficult to connect i and iii directly.

- i ⟹ ii: Construct a sequence of distinct terms recursively. At the inductive step you will need to show that you can pick an element of A that has not already been picked. Assume you cannot, and show that then A is finite—that is, it satisfies the definition of a finite set.

- ii ⟹ i: Proceed by contradiction. Use a modification of the proof technique used for Theorem 7.2.3.

"I'M BEGINNING TO UNDERSTAND ETERNITY, BUT INFINITY IS STILL BEYOND ME."

Figure 7.2

- ii $\Rightarrow$ iii: Let (a_i) be a sequence of distinct terms in A. Let $B = A \setminus \{a_i\}_{i=1}^{\infty}$. Show that A can be put into one-to-one correspondence with $\{a_2, a_3, a_4, \ldots\} \cup B$.

- iii $\Rightarrow$ ii: Suppose that A can be put into one-to-one correspondence with its proper subset C. Let $f : A \to C$ be that one-to-one correspondence. Let $a_0 \in A \setminus C$. Construct a sequence by iterating f starting at a_0. Show that you get a sequence of distinct terms.   □

7.2.6   COROLLARY

A set is infinite if and only if it has an infinite subset.   □

The following corollary tells us that there are no infinite sets that are "smaller" than $\mathbb{N}$.

7.2.7   COROLLARY

Let C be any infinite set. Then card $\mathbb{N} \leq$ card C.
   (*Hint:* Remember that a sequence is a function.)   □

7.3   Countable Sets

The smallest possible infinite sets (those sets that have the same cardinality as $\mathbb{N}$) have a special name.

7.3.1   DEFINITION

Let A be a set. A is said to be **denumerable** or **countably infinite** if A and $\mathbb{N}$ have the same cardinality. A is said to be **countable** if it is finite or denumerable.

7.3.2   EXAMPLE

You have already seen several examples.

1. Clearly $\mathbb{N}$ is denumerable.

2. Galileo showed that the set of perfect squares is countable.

3. In Exercise 7.1.2 you showed that the even natural numbers and the integers are countably infinite sets.   ■

> If A is a countably infinite set, the elements of A can be listed as a sequence of distinct terms:
>
> $$A = \{a_1, a_2, a_3, \ldots\}.$$
>
> This is a very useful property. You will often need to use it when proving theorems about countable sets.
>   Conversely, if you want to show that a set is countable, one possibility is to show that its elements can be listed as a sequence of distinct terms.

What general sorts of things can we say about countable sets?

7.3.3   THEOREM

Let C be a countable set and B be any set. If $f : C \to B$ is a one-to-one correspondence, then B is countable.   □

7.3.4   THEOREM

Every subset C of a countable set A is countable.

(*Hint:* Notice that in the case where A is finite this is the contrapositive of Corollary 7.2.6.)   □

7.3.5   THEOREM

Let $f : \mathbb{N} \to X$ be an onto function. Then X is countable.   □

> This last theorem is interesting for a couple of reasons. Intuitively, it says that if the elements of $\mathbb{N}$ can entirely "cover" the elements of X, then X can be no larger than $\mathbb{N}$. Practically speaking, it tells us that we can show a set X is countable by constructing a function from $\mathbb{N}$ onto X; we don't have to worry about making it one-to-one!

All the examples of countable sets that we have seen so far share one distinctive feature: Their elements are "separated." If we consider $\mathbb{Z}$, for instance, each element has an immediate successor and predecessor. When we planned our strategies for constructing functions from $\mathbb{N}$ to the sets, we exploited this property—we used "next" elements to pick the elements of the set one at a time without missing any. If we were to try to tackle a set, such as the rational numbers, where there are no immediate successors or predecessors, things would not be quite so easy. Having picked one rational number, when we pick another we will inevitably miss many that fall between. There is no guarantee that we will ever be able to go back and pick all these up. In fact, the task seems hopeless. Surprisingly, it is not! The solution rests, as it did for the integers, on organizing the elements in the set and planning a "course of action."

7.3.6 THEOREM

The set of rational numbers is countable.

(*Hint:* Using the chart shown in Figure 7.3 to "picture" the rationals should help you plan your strategy.) □

The following set-theoretic lemma and the exercise following it should be useful in the proof of Theorem 7.3.10.

7.3.7 LEMMA

Let $\{A_n\}_{n=1}^{\infty}$ be a collection of sets. Let $B_1 = A_1$, and for each natural number $n > 1$ let

$$B_n = A_n \setminus \bigcup_{i=1}^{n-1} A_i.$$

Then:

1. The collection $\{B_n\}_{n=1}^{\infty}$ is pairwise disjoint. (*Hint:* Use set algebra!)

2. $\displaystyle\bigcup_{n=1}^{\infty} A_n = \bigcup_{n=1}^{\infty} B_n.$ □

	0	1	-1	2	-2	3	-3	4	-4	$\cdots$
1	$\frac{0}{1}$	$\frac{1}{1}$	$-\frac{1}{1}$	$\frac{2}{1}$	$-\frac{2}{1}$	$\frac{3}{1}$	$-\frac{3}{1}$	$\frac{4}{1}$	$-\frac{4}{1}$	$\cdots$
2	$\frac{0}{2}$	$\frac{1}{2}$	$-\frac{1}{2}$	$\frac{2}{2}$	$-\frac{2}{2}$	$\frac{3}{2}$	$-\frac{3}{2}$	$\frac{4}{2}$	$-\frac{4}{2}$	$\cdots$
3	$\frac{0}{3}$	$\frac{1}{3}$	$-\frac{1}{3}$	$\frac{2}{3}$	$-\frac{2}{3}$	$\frac{3}{3}$	$-\frac{3}{3}$	$\frac{4}{3}$	$-\frac{4}{3}$	$\cdots$
4	$\frac{0}{4}$	$\frac{1}{4}$	$-\frac{1}{4}$	$\frac{2}{4}$	$-\frac{2}{4}$	$\frac{3}{4}$	$-\frac{3}{4}$	$\frac{4}{4}$	$-\frac{4}{4}$	$\cdots$
5	$\frac{0}{5}$	$\frac{1}{5}$	$-\frac{1}{5}$	$\frac{2}{5}$	$-\frac{2}{5}$	$\frac{3}{5}$	$-\frac{3}{5}$	$\frac{4}{5}$	$-\frac{4}{5}$	$\cdots$
$\vdots$	$\vdots$	$\vdots$	$\vdots$	$\vdots$	$\vdots$	$\vdots$	$\vdots$	$\vdots$	$\vdots$	

Figure 7.3 The rational numbers

7.3.8 EXERCISE

State and prove an analog of Lemma 7.3.7 for a finite collection of sets.   □

7.3.9 EXERCISE

Show that the union of two countable sets is countable.   □

This is a special case of the more general:

7.3.10 THEOREM

Every countable union of countable sets is countable.

□

A *countable union* is a union of countably many sets. That is, it is either a union of finitely many sets or of denumerably many sets:

$$\bigcup_{i=1}^{n} A_i \quad \text{or} \quad \bigcup_{i=1}^{\infty} A_i.$$

Prove Theorem 7.3.10 in several stages.

1. Every finite union of countable sets is countable. (*Hint:* Proceed by induction on the number of sets starting with a base case of 2.)
2. Every denumerable union of finite sets is denumerable.
3. Every denumerable union of denumerable sets is denumerable. (*Hint:* A chart similar to the one shown in Figure 7.3 should help.)
4. Finally, put it all together to derive the result of Theorem 7.3.10.

7.3.11 EXERCISE

Use Theorem 7.3.10 to show that:

1. If A and B are countable sets, then $A \times B$ is a countable set.
2. The set of all finite sequences in the set $S = \{a, b\}$ is a countable set. (A finite sequence in $S = \{a, b\}$ is a finite "string" of a's and b's. For instance:
   - a, b, b, a is a sequence of length 4.
   - a, a, a, a, a is a sequence of length 5.
   - b, a, a, b, a, a is a sequence of length 6.)   □

7.4   Beyond Countability

Recall the first scenario that was discussed in Section 7.2. We pictured our set as a bunch of things in a basket. We drew those things out one at a time and discarded them as we went. In the case of infinite sets this process never terminates, and we end up with a sequence of distinct terms. The question is, will the sequence we draw out exhaust the set? That is, will each member of the set be one of the elements in the sequence? Well, on the face of it, the answer is obviously no. If our set were $\mathbb{N}$, we could draw out the sequence of even numbers, leaving infinitely many elements behind. However, in the case of $\mathbb{N}$ it is clear that, by adjusting our strategy for taking out elements, we can certainly include all the natural numbers in our sequence. Thus we see that the interesting question is not whether a particular sequence we might choose will exhaust the set, but whether it is possible to *adopt a sufficiently clever strategy* for drawing out elements so that we actually get all the set in the process. A set is countably infinite if and only if we can find a strategy that will allow us to exhaust the set.

7.4.1 EXERCISE

Look back at Definition 7.1.7. How does the idea of adopting a "clever strategy" relate to this definition? How would you phrase the ability to find such a strategy in functional terms? Suppose we could find a set for which no such strategy existed. What would that say about its relationship to $\mathbb{N}$? □

After he proved that the rationals were countable, Cantor began to think that all sets were countable. No one was more amazed than he was when he managed to prove otherwise.

7.4.2 DEFINITION

A set that is not countable is said to be **uncountable**.

Cantor devised a simple, ingenious, and beautiful proof that tells us there are uncountable sets. Here it is.

7.4.3 THEOREM (Cantor's diagonalization argument)

The set of all real numbers between 0 and 1 is uncountable.

Proof. We will prove that there exists no onto function from $\mathbb{N}$ to $(0, 1)$. (In this case, of course, there can be no one-to-one correspondence between them.)

Let $f : \mathbb{N} \to (0, 1)$ be any function. Because every element of $(0, 1)$ can be written in decimal form, we denote $f(n)$ as follows:

$$f(n) = .a_{1n} a_{2n} a_{3n} \cdots,$$

where a_{1n} is the first digit in the decimal expansion, a_{2n} is the second digit, and so forth. To avoid ambiguity, where there is a choice, we take the terminating rather than the repeating expansion.[2]

Claim: There is a point $b \in (0, 1)$ that has nothing mapping to it. (That is, f is not onto.)

Let $b = .b_1 b_2 b_3 \cdots$, where

$$b_i = \begin{cases} 2 & \text{if } a_{ii} \neq 2, \\ 3 & \text{if } a_{ii} = 2. \end{cases}$$

---

[2] For example, we take $0.20000\ldots$ rather than $0.1999\ldots$, which represent the same number.

Note that since no digit in the expansion of b is a 9, b has no terminating expansion. So b is expanded in decimal form according to the criterion we set above.

We must now show that $f(n) \neq b$ for any n. First we note that two numbers expanded unambiguously in decimal form differ from one another if and only if they differ in at least one decimal place. The nth decimal of $f(n)$ is a_{nn}. The nth decimal of b was deliberately chosen to be different from a_{nn}, so (for each $n \in \mathbb{N}$) $f(n)$ differs from b at least in the nth decimal place. We conclude that f is not onto. Since f was an arbitrary function, there are no one-to-one correspondences between $\mathbb{N}$ and $(0, 1)$. ■

7.4.4 COROLLARY

The set $\mathbb{R}$ of real numbers is uncountable. □

7.4.5 COROLLARY

Let a and $b \in \mathbb{R}$ with $a < b$. Then (a, b), $[a, b]$, $(a, b]$, and $[a, b)$ are all uncountable. □

7.4.6 COROLLARY

The set of irrational numbers is uncountable. □

We now know that $\mathbb{R}$ is "bigger," in some real sense, than the natural numbers and even the rationals. Are there sets that are bigger than $\mathbb{R}$? Yes. Cantor was able to show that no matter how big a set is, there is always a bigger set.

An exercise will be helpful before the proof.

7.4.7 EXERCISE

Let $A = \{a, b, c, d, e, f\}$. Define $g : A \to \mathcal{P}(A)$ by

$$g(a) = \{a, d, e\},$$
$$g(b) = \{c, f\},$$
$$g(c) = \emptyset,$$
$$g(d) = \{b, c, d, e\},$$
$$g(e) = \{a, b, c, d, e, f\},$$
$$g(f) = \{b\}.$$

1. Find the set $J = \{x \in A : x \notin g(x)\}$.
2. Is $J \in \mathcal{R}an(g)$?
3. The result you found in part 2 is not coincidental. Ask yourself, if some element $x \in A$ *did* map to J, would x be an element of J or not? □

7.4.8 THEOREM (Generalized Cantor diagonalization argument)

Let A be a set. Then card $A <$ card $\mathcal{P}(A)$.

(*Hint:* As in Cantor's diagonalization argument, we show that no function $g :$ $A \to \mathcal{P}(A)$ can be onto. Let g be any function from A into $\mathcal{P}(A)$. Let $J = \{x \in A : x \notin g(x)\}$. Show that J is not in the range of g.)   □

Theorem 7.4.3 says that there is an uncountable set. Theorem 7.4.8 says that given any set (no matter how big) there is a set with larger cardinality. In fact, for any set A:

$$\text{card } A < \text{card } \mathcal{P}(A) < \text{card } \mathcal{P}(\mathcal{P}(A)) < \cdots .$$

Since we originally thought of cardinality as a way of talking about "how large" an infinite set is, this should tell you that (in some reasonable sense) there are many "sizes of infinity."

7.4.9   PROBLEM

Prove that $\mathcal{P}(\mathbb{N})$ has the same cardinality as the set of all sequences of 0's and 1's. (To help with the intuition, think about 0 as meaning "no" and 1 as meaning "yes.")   □

This is a special case of a more general theorem.

7.4.10   PROBLEM

Let K be a set. We define a function $\chi_K : K \to \{0, 1\}$ as follows:

$$\chi_K(x) = \begin{cases} 1 & \text{if } x \in K, \\ 0 & \text{if } x \notin K. \end{cases}$$

χ_K is called the **characteristic function of** K. Let X be any set. Show that the set of characteristic functions of subsets of X has the same cardinality as $\mathcal{P}(X)$.   □

7.5   Comparing Cardinalities

So far we have come tantalizingly close to saying that we can compare the size of any two sets by using functions. We talk about card A as though we could actually assign a "number" of elements to A. We use the symbol $\leq$ to compare cardinalities, making us think that we are actually comparing those numbers, just as we can compare real numbers. It is now time to place this all into the context of orderings.
Here is the plan:

> To keep ourselves safely away from Russell's paradox, we will work inside a fixed set X. All sets I refer to in this section are assumed to be subsets of X.

Step I. I will at last define the cardinality of a set.

Step II. We will then prove that the set of cardinalities can be partially ordered. To do so, of course, we have to define the relation $\leq$ (which will be closely related to the relation we have been discussing all along) and then prove that:

$$\le \; is \; reflexive, \quad \le \; is \; transitive, \quad and \quad \le \; is \; antisymmetric.$$

(The antisymmetry of the relation—the famous Schroeder-Bernstein theorem—is quite deep and is an important result in its own right.)

Though $\le$ is, in fact, a total ordering, proving this fact is beyond the scope of this book. To do so, we would have to delve more deeply into the intricacies of set theory than we can do here (or even in the technical appendices).

You showed in Theorem 7.1.3 that the subsets of X can be partitioned according to their cardinality. This tells us that given a set A, we can characterize all subsets of X that have the *same* cardinality as A: Just look at the equivalence class of A under this equivalence relation. When thinking about cardinality, we cannot distinguish between different elements of an equivalence class; thus when we talk about card A, we are really indirectly referring to all elements in a given equivalence class. This gives us our first real clue to the mathematical definition of the cardinality of a set.

7.5.1 DEFINITION

The **cardinality** of A is the equivalence class of all subsets of X with the same cardinality as A.

Previously, "card A = card B" and "card $A \le$ card B" were merely symbolic ways of indicating functional relationships between the sets A and B; "card A" had no meaning by itself. Now that we have defined card A, the expression card A = card B takes on a new meaning; it says that two equivalence classes are equal. It is important to make sure that our old and new interpretations are commensurate.

7.5.2 EXERCISE

Let A and B be sets. Show that card A = card B in the sense of Definition 7.1.1 if and only if card A and card B are the same equivalence class.

(*Hint:* You have really proved this already. Look at Lemma 4.3.20.)   □

We also wish to interpret card $A \le$ card B as a relation between the *equivalence classes* card A and card B, not merely between the sets A and B.

7.5.3 DEFINITION

Let card A and card B be the cardinalities of subsets A and B of X. Then we say that card $A \le$ card B if there exists a one-to-one function from A into B.

The form of Definition 7.5.3 raises the crucial issue of well-definedness.[3] A relation is well-defined if there is a definite procedure for determining whether two elements are related and that procedure *always yields the same answer.* To see what happens in this case, let us examine the procedure that is prescribed by Definition 7.5.3.

Suppose we are asked whether card A is related to card B. Definition 7.5.3 tells us that if there is a one-to-one function from A into B, then card $A \leq$ card B. (So far, so good.) Now suppose that A and C have the same cardinality and B and D have the same cardinality. Then card A and card C are the same equivalence class referred to by different names, as are card B and card D. Asking whether card A is related to card B is the *same* as asking whether card C is related to card D. But now Definition 7.5.3 asks us whether there is a one-to-one function from C into D. Will we get the same answer? If not, the relation is not well-defined. Fortunately for us, $\leq$ *is* well-defined, and this follows from our next theorem.

7.5.4   THEOREM

Suppose that card $A =$ card C and card $B =$ card D. Then there is a one-to-one function from A into B if and only if there is a one-to-one function from C into D. Thus $\leq$ is a well-defined relation.   □

Now that we have our relation, we can proceed to show that it is a partial order.

7.5.5   THEOREM

Prove that the relation $\leq$ on the set of cardinalities of subsets of X is reflexive and transitive.   □

Suppose A is a finite set and (as before) we denote the number of elements of A by $\#A$. Our statement that $\leq$ is antisymmetric is analogous to saying (about finite sets) that if $\#A \leq \#B$ and if $\#B \leq \#A$, then A and B have the same number of elements. It seems deceptively simple, yet the proof requires a good deal of insight into the problem and ingenuity besides. Included is a fairly detailed outline of the proof; the details are left for you to fill in. Before you read through the proof sketch, however, take some time to think about the statement of the theorem. Your first instinct will tell you that it should be *easy* to prove. Is it really necessary to go to all this trouble? Don't proceed until you have at least a vague sense of why this is hard.

7.5.6   THEOREM (Schroeder-Bernstein)

If the card $A \leq$ card B and card $B \leq$ card A, then card $A =$ card B.

---

[3] If you studied Chapter 6 on number theory, you have already encountered the question of well-definedness in the discussion of addition and multiplication modulo n. However, to keep this chapter independent of that material, I will not assume familiarity with Chapter 6 here.

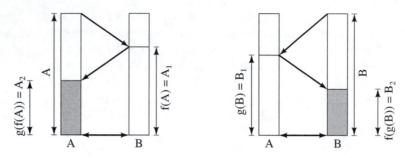

Figure 7.4   Schroeder-Bernstein 1

Formally speaking, this says that given two sets A and B; if there is a one-to-one function f from A into B and a one-to-one function g from B into A, then there is a one-to-one correspondence h between A and B.

Proof Sketch.

Step I. Define $A_0 = A$ and $B_0 = B$; $A_1 = f(A_0)$ and $B_1 = g(B_0)$. Then for each $n \in \mathbb{N}$ define:

$$A_{2n} = g(A_{2n-1}), \qquad A_{2n+1} = f(A_{2n}),$$
$$B_{2n} = f(B_{2n-1}), \qquad B_{2n+1} = g(B_{2n}).$$

(These sets are illustrated in Figure 7.4.)

Step II. Prove that

$$A_0 \supseteq B_1 \supseteq A_2 \supseteq B_3 \supseteq A_4 \supseteq B_5 \cdots$$

and

$$B_0 \supseteq A_1 \supseteq B_2 \supseteq A_3 \supseteq B_4 \supseteq A_5 \cdots.$$

Step III. Use Step II to deduce that

$$\bigcap_{i=0}^{\infty} A_{2i} = \bigcap_{i=0}^{\infty} B_{2i+1} \quad \text{and} \quad \bigcap_{i=0}^{\infty} B_{2i} = \bigcap_{i=0}^{\infty} A_{2i+1}.$$

Define

$$A_{\infty} = \bigcap_{i=0}^{\infty} A_{2i} = \bigcap_{i=0}^{\infty} B_{2i+1} \quad \text{and} \quad B_{\infty} = \bigcap_{i=0}^{\infty} B_{2i} = \bigcap_{i=0}^{\infty} A_{2i+1}.$$

Step IV. Show that the set

$$\{A_{\infty}, (A_0 \setminus B_1), (B_1 \setminus A_2), (A_2 \setminus B_3) \ldots\}$$

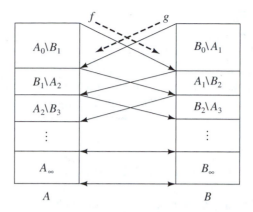

Figure 7.5 Schroeder-Bernstein 2

partitions the set A, and that the set

$$\{B_\infty, (B_0 \setminus A_1), (A_1 \setminus B_2), (B_2 \setminus A_3) \ldots\}$$

partitions the set B. (These partitions are shown in Figure 7.5.)

Step V. Define $h : A \to B$ by

$$h(x) = \begin{cases} f(x) & \text{if } x \in A_{2i} \setminus B_{2i+1} \text{ for some } i \in \mathbb{N} \cup \{0\}, \\ g^{-1}(x) & \text{if } x \in B_{2i+1} \setminus A_{2(i+1)} \text{ for some } i \in \mathbb{N} \cup \{0\}, \\ f(x) & \text{if } x \in A_\infty. \end{cases}$$

Step VI. Prove that h is a function.

Step VII. Prove that h is one-to-one and onto. □

Thus $\le$ is a partial order on the set of cardinalities of subsets of X.

7.5.7 EXERCISE

Use the Schroeder-Bernstein theorem to prove that if $A \subseteq B \subseteq C$ and card $A =$ card C, then A, B, and C all have the same cardinality. □

This leaves us yet with the question of whether the cardinalities of any two sets may be compared, that is, whether $\le$ is a total order on the set of cardinalities. According to Theorem 4.2.6, we would have to prove the following theorem.

7.5.8 THEOREM

Let A and B be subsets of X. Exactly one of the following is true:

- card $A <$ card B.

- card $A =$ card B.
- card $A >$ card B.   □

I have already said that the proof of Theorem 7.5.8 is beyond the scope of this book. If you would like to know more, you should consult a work on axiomatic set theory. My favorite is a very well-written volume called *Introduction to Set Theory* by Karel Hrbacek and Thomas Jech (Dekker, 1984), but there are many good ones.

7.6   The Continuum Hypothesis

We know that the smallest infinite sets are the denumerable sets. We also know that given any infinite set, we can find a set with larger cardinality by taking its power set. In particular, card $\mathbb{N} <$ card $\mathcal{P}(\mathbb{N})$.

Question How *much* bigger is card $\mathcal{P}(\mathbb{N})$ than card $\mathbb{N}$? Are there cardinalities in between? How do we know?

As is our custom when considering new questions about the size of infinite sets, let us consider the analogous question about finite sets. Suppose we have a set with three elements. Then its power set has eight elements. Are there sets with cardinalities between 3 and 8? Sure—$\{1, 2, 3, 4, 5, 6\}$, for instance. In fact, given any set with n elements, its power set has 2^n elements, and if $n > 1$, we can always find cardinalities between n and 2^n. Thus we might expect there to be sets that are larger than $\mathbb{N}$ and smaller than $\mathcal{P}(\mathbb{N})$.

Let us consider for a moment what else we know. Cantor proved that card $\mathbb{N} <$ card$(0, 1)$. It is not too difficult to show that $\mathbb{R}$ has the same cardinality as $(0, 1)$.

So we currently see the hierarchy in cardinalities shown in Figure 7.6. We might be tempted to compare card $\mathcal{P}(\mathbb{N})$ with card $\mathbb{R}$. If we did we would find the following theorem.

7.6.1   THEOREM

$\mathcal{P}(\mathbb{N})$ and $\mathbb{R}$ have the same cardinality.

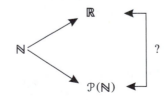

Figure 7.6   Which is bigger, $\mathbb{R}$ or $\mathcal{P}(\mathbb{N})$?

Hint for proving Theorem 7.6.1: Express each element of (0, 1) in a binary expansion and thus as a sequence of 0's and 1's. (To avoid ambiguity, take the terminating rather than the nonterminating expansion where applicable.) Now we have a natural one-to-one association between elements of (0, 1) and the subsets of $\mathbb{N}$.

As in Problem 7.4.9, identify each subset of $\mathbb{N}$ with its characteristic function $\chi_{\mathbb{N}} : \mathbb{N} \to \{0, 1\}$: a sequence of 0's and 1's that is 0 at points that are not contained in the set and 1 at points that are contained in the set. Map the sequence $a_1, a_2, a_3, \ldots$ to the number in (0, 1) with *decimal* expansion $.a_1 a_2 a_3 \ldots$. Show that this is a one-to-one function from $\mathbb{N}$ into (0, 1).

Finally, apply the Schroeder-Bernstein theorem.

□

> **Binary Expansions:** Instead of writing a number as a sum of powers of ten, write it as a sum of powers of two. For instance, $12.421875 = (1 \times 2^3) + (1 \times 2^2) + (0 \times 2^1) + (0 \times 2^{-1}) + (1 \times 2^{-2}) + (1 \times 2^{-3}) + (0 \times 2^{-4}) + (1 \times 2^{-5})$. Thus the binary expansion of 12.421875 is 1100.011011. It takes some work to show that all the real numbers must have a binary expansion, but it can be done.

Given this, it is not too hard to prove that the following statements are equivalent.

i. There exists a set K such that card $\mathbb{N}$ < card K < card $\mathcal{P}(\mathbb{N})$, and

ii. There is a subset S of $\mathbb{R}$ such that card $\mathbb{N}$ < card S < card $\mathbb{R}$.

So asking whether there are cardinalities between card $\mathbb{N}$ and card $\mathcal{P}(\mathbb{N})$ is equivalent to asking whether there is a set of real numbers that is bigger than $\mathbb{N}$ but smaller than $\mathbb{R}$. Is this plausible? What would such a set look like? Remember, the rationals are denumerable, yet every interval has the cardinality of $\mathbb{R}$. As it turns out, the irrationals have the same cardinality as $\mathbb{R}$, too. In fact, all known ways of producing subsets of $\mathbb{R}$ yield either countable sets or sets with the same cardinality as $\mathbb{R}$. Does this perhaps shed some doubt on the issue of whether there are sets with cardinality bigger than $\mathbb{N}$ and smaller than $\mathcal{P}(\mathbb{N})$?

Cantor originally conjectured that there are no sets (in cardinality) "between" $\mathbb{N}$ and $\mathbb{R}$. This conjecture is called the *continuum hypothesis*.

The Continuum Hypothesis. There does not exist a set S so that

> $\mathbb{R}$ is sometimes called the continuum. Thus we talk about "the cardinality of the continuum."

card $\mathbb{N}$ < card S < card $\mathbb{R}$.

The famous question remained unanswered for more than seven decades! In the late 1930s, Kurt Gödel proved that the continuum hypothesis is consistent with the standard axioms of set theory. That is, that it is not possible, on the basis of the axioms, to prove it false. In the early 1960s, P. J. Cohen proved that the negation of the continuum hypothesis is *also* consistent with the axioms of set theory. That is, it is impossible to prove the continuum hypothesis to be true on the basis of the axioms! What we have here is a statement that can neither be proved true nor false based on the standard axioms of set theory.

All this might remind you about the controversy surrounding Euclid's fifth postulate for geometry. (See Section 0.1 for a brief description.) That controversy ended in the early nineteenth century with the discovery of alternative (but entirely consistent) geometries that were based on the acceptance of axioms contradictory to the fifth postulate. Are there *other* statements out there that we will be unable to prove or disprove based on our standard axioms? Yes, in fact, Kurt Gödel proved that no finite set of axioms (strong enough to yield the elementary facts about arithmetic) will ever be able to answer every mathematical question. *It does not matter how thorough or clever we are—we can never choose a complete set of axioms!* For a very nice nontechnical explanation of Gödel's theorem, see *Gödel's Proof* by Nagel and Newman (New York University Press, 1958). There are more rigorous mathematical treatments. Try, for instance, *Gödel's Incompleteness Theorems* by Raymond Smullyan (Oxford University Press, 1992).

> A statement that can neither be proved nor disproved on the basis of a set of axioms is said to be *independent* of the axioms. We are free to accept it as another axiom, or accept its negation. Either assumption is completely consistent. The two alternatives will lead to mutually contradictory but entirely self-consistent parallel mathematical universes.

■ PROBLEMS

1. Show that for all real numbers a and b with $a < b$, (a, b) has the same cardinality as $\mathbb{R}$.

2. Suppose A and B are sets such that card $A \leq$ card B. Prove that there exists a set $C \subseteq B$ such that card $C =$ card A.

3. Suppose that A, B, and C are sets such that card $A <$ card B and card $A =$ card C. Prove that card $C <$ card B.

4. Let A, B, and C be sets. Suppose that card $A <$ card B and card $B \leq$ card C. Prove that card $A <$ card C. (A parallel statement holds if card $A \leq$ card B and card $B <$ card C.)

5. Let K be any set and let $\mathcal{F}$ be the set of all functions with domain K. Then card $K <$ card $\mathcal{F}$.

6. Let a and b be real numbers with $a < b$. Prove that (a, b) has the same cardinality as $[a, b]$.

7. Prove that the set of irrational numbers has the same cardinality as the set of real numbers.

■ QUESTIONS TO PONDER

1. Let A and B be finite sets. Prove that the following statements are equivalent:

   (a) There is a one-to-one function from A into B that is not onto.

(b) There is a one-to-one function from A into B but there is no one-to-one correspondence between A and B.

2. Prove the following generalization of Theorem 7.3.5:

   Let X and Y be sets. If there exists a function $f : X \to Y$ that is onto, then card $Y \leq$ card X.

3. On page 175 I described binary expansions of real numbers and said, "It takes some work to show that all the real numbers must have a binary expansion, but it can be done." *How* might it be done? Think about what sorts of mathematical ideas would be required for a proof of this fact. Can you prove it?

4. **Definition:** A totally ordered set A with at least two elements is said to be **dense in itself** if given any two elements x and y in A with $x < y$, there exists an element $z \in A$ such that $x < z < y$.

   Prove the following theorem:

   Any countable, dense in itself, totally ordered set with no greatest element and no least element is order isomorphic to the rational numbers under the usual ordering of real numbers.

   A proof that a mathematical object is unique up to isomorphism is called a *complete characterization*.

5. If we have two disjoint finite sets A and B, then we know that

$$\#A + \#B = \#(A \cup B).$$

   Suppose we have two disjoint *infinite* sets A and B. Suppose that card $A \leq$ card B. What can you say about card$(A \cup B)$? Can it be larger than card B?

6. I have said that proving Theorem 7.5.8 is well beyond the scope of this book. Indeed, it is a deep theorem. Play with the idea. Try to uncover the central difficulty. Why is it so hard to prove?

8 The Real Numbers

8.1 Constructing the Axioms

The real numbers are without a doubt the fundamental mathematical structure. Our purpose in this chapter is to introduce the axioms for the real number system. As we do this, we will discuss the desirability and necessity of the axioms that we choose. *Desirability:* Our axioms should describe the familiar properties of the real number system that we want to be sure to incorporate in our study of them. *Necessity:* The real number system has many properties—it is possible to list hundreds of "facts" about the real numbers—but is it necessary to accept them all as axioms? Clearly not. Many of these facts can be proved from other facts. We will want to choose for our axioms a few crucial statements from which other facts about the real number system can be proved.

We will start with a set $\mathbb{R}$, whose elements we will call *real numbers*. Our study of $\mathbb{R}$ depends, first and foremost, on set theory; but $\mathbb{R}$ clearly has a structure that goes beyond a simple description of its contents. The properties of arithmetic, order, and so forth must be specified by means of axioms.

8.2 Arithmetic

Small children begin their study of mathematics with the arithmetic operations of addition, subtraction, multiplication, and division. We clearly want to have these operations on real numbers at our disposal.

Since we can think of $a - b$ as $a + -b$ and $\frac{a}{b}$ as $a \cdot \frac{1}{b}$, our axioms need only discuss addition and multiplication. These axioms must, however, tell us what is meant by $-b$ and $\frac{1}{b}$. ($-b$ is technically called the *additive inverse* of b. $\frac{1}{b}$ is called the *multiplicative inverse* of b.) A discussion of inverses presupposes the existence of additive and multiplicative identities. Thus our axioms will discuss addition and multiplication, incorporating identities and inverses for each.

As we will see later in the section, it is important to say something about the way that addition and multiplication interact with one another, so our axioms must say something about this as well.

AXIOM I (The Field Axioms)

Let $\mathbb{R}$ be a set. We assume the existence of two binary operations on $\mathbb{R}$ that we will call $+$ (addition) and $\cdot$ (multiplication). We assume that these binary operations have the following properties:

- **Addition and multiplication are commutative operations.** For all real numbers x and y,

$$x + y = y + x \quad \text{and} \quad x \cdot y = y \cdot x.$$

- **Addition and multiplication are associative operations.** For all real numbers x, y, and z,

$$(x + y) + z = x + (y + z) \quad \text{and} \quad (xy)z = x(yz),$$

where the parentheses, as usual, indicate which operations are performed first.

- **There exist additive and multiplicative identities.** There exists a real number that we will call 0 (zero) such that for all real numbers x, $x + 0 = x$. There exists a real number distinct from zero that we will call 1 (one) such that for all real numbers x, $x \cdot 1 = x$.

> Notice that I have suppressed the symbol $\cdot$ in the expression
> $$x \cdot (y \cdot z) = (x \cdot y) \cdot z$$
> and represented multiplication in the conventional way by concatenating the elements we wish to multiply. I will continue to do this in cases where no confusion can arise.

- **There exist additive and multiplicative inverses.** For each real number x there exists a real number that we will call $-x$ (minus x) such that $x + (-x) = 0$. For each real number x *except zero* there exists a real number that we will call $\frac{1}{x}$ (one over x) such that $x \cdot \frac{1}{x} = 1$.

- **Multiplication distributes over addition.** For all real numbers x, y, and z,

$$x(y + z) = xy + xz.$$

(We follow the usual convention that, in the absence of parentheses, multiplications are performed before additions.) □

In effect, $\mathbb{R}$ is an algebraic structure called a **field.** Though we will not dwell on this terminology here, a field is any algebraic structure that satisfies the field axioms that were just given.

As suggested above, we now *define* subtraction and division.

8.2.1   DEFINITION (Subtraction and division)

Let x and y be real numbers. We define binary operations $-$ and $\div$ on $\mathbb{R}$ as follows:

$$x - y = x + (-y) \quad \text{and} \quad x \div y = x \cdot \frac{1}{y}.$$

(Following the convention, I will usually write $\frac{x}{y}$ instead of $x \div y$.)[1]

The following theorem says that additive and multiplicative inverses are unique.

8.2.2 THEOREM (Uniqueness of inverses)

Let x be a real number. Prove that there exists only one real number y such that $x + y = 0$. Similarly, if $x \neq 0$, prove that there exists only one real number y such that $x \cdot y = 1$. □

8.2.3 EXERCISE

Use Theorem 8.2.2 to prove that for all $x \in \mathbb{R}$, $-(-x) = x$, and for all $x \in \mathbb{R}$ such that $x \neq 0$, $\frac{1}{1/x} = x$. □

Consider the following list of statements about the real numbers. Our previous experience tells us that they are true, but they are not explicitly assumed in the field axioms:

I. For all real numbers x, $x \cdot 0 = 0$. (Zero has multiplicative properties as well as additive properties.)

II. For all real numbers x, $-x = (-1) \cdot x$. (Think carefully about this. Many students have trouble with this one because they think that this is, in fact, the *definition* of $-x$. But the description of $-x$ in the field axioms never mentions multiplication at all!)

III. For all nonzero numbers a, $a \neq -a$.

IV. $\mathbb{R}$ is an infinite set.

V. For all positive real numbers a there exists a number x so that $x^2 = a$.

8.2.4 THEOREM

For all real numbers x, $x \cdot 0 = 0$.

(*Hint*: Since this theorem discusses multiplication and the only thing we know about 0 is that it behaves in a certain way with respect to addition, you will clearly have to use the only axiom that talks about the connection between addition and multiplication.) □

8.2.5 THEOREM

For all real numbers x, $-x = (-1) \cdot x$.

(*Hint:* See the hint for Theorem 8.2.4.) □

---

[1] Well, OK, I lied. $\div$ is not, strictly speaking, a binary operation on $\mathbb{R}$ because it is not defined on all pairs of real numbers. Zero has no multiplicative inverse, so $x \div 0$ does not make any sense—this fits in with our usual notion that we cannot divide by zero. With this exception, however, division *is* an operation that allows us to take two real numbers and obtain from them a third. Division violates the letter of the definition but not the spirit, so I take the liberty of calling it a binary operation in this case and assuage my conscience by confessing.

The following problem will tell you that there are mathematical objects that are *not* $\mathbb{R}$ but that nevertheless satisfy the field axioms.

8.2.6　PROBLEM

Let A be a set containing two elements: the set of even integers and the set of odd integers. (We will call the set of even integers E and the set of odd integers O.) We define two binary operations $\oplus$ and $\otimes$ on A in the following intuitive way. Since the sum of two even integers is even, we define $E \oplus E$ to be E. Similarly, since the product of an even integer and an odd integer is even, we define $O \otimes E$ to be E. The remaining sums and products are defined analogously and are summarized in the following tables.[2]

$\oplus$	E	O
E	E	O
O	O	E

and

$\otimes$	E	O
E	E	E
O	E	O

1. Verify that A together with $\oplus$ and $\otimes$ satisfies the field axioms.
2. Identify "0," "1," and "-1."　　　　　□

In Problem 8.2.6 you saw a mathematical structure composed of a set and two binary operations that satisfies all of the axioms we have adopted to date. This clearly means that, assuming only the field axioms, any theorem that can be proved about the real numbers will also be true of the example given in Problem 8.2.6. In particular, since A is finite, we will not be able to prove statement IV using nothing but the field axioms. Furthermore, since $1 = -1$ in the example, we have no hope of proving statement III.

What about statement V? It talks about our ability to take square roots of positive numbers. Of course, we understand intuitively what is meant by "positive numbers," but as yet we have no mathematical definition for them. We will need a way of getting at the positive numbers in order for statement V to make any sense.

How might we define the positive numbers? Let's see . . . the positive numbers are those that are greater than zero. Unfortunately, this definition requires an order on $\mathbb{R}$, something that we do not yet have. On the other hand, if we know how to recognize positive numbers when we see them, we can use this knowledge to define an order on $\mathbb{R}$ by saying that $x < y$ if $y - x$ is positive. If we have an order on $\mathbb{R}$, we can define "positive." If we know what positive means, we can define an order on $\mathbb{R}$. The two ideas are mathematically equivalent.

---

[2] Of course, we could have defined these operations on any two-point set and the mathematical result would have been the same. The "even/odd" scheme merely provides a motivation and a way to remember the results.

8.3   Order

Statements III and V in Section 8.2 each touch on the question of positivity in their own way. The following axiom gives us the positive numbers.

AXIOM II (The order axiom)

There exists a subset $\mathbb{R}^+$ of $\mathbb{R}$ such that

1. $\mathbb{R}^+$ is closed under addition and multiplication. That is, for all x, $y \in \mathbb{R}^+$,

$$x + y \in \mathbb{R}^+ \quad \text{and} \quad x \cdot y \in \mathbb{R}^+.$$

2. For all real numbers a, one and only one of the following holds:
   - $a \in \mathbb{R}^+$.
   - $a = 0$.
   - $-a \in \mathbb{R}^+$.

We call the elements of $\mathbb{R}^+$ **positive numbers.** We call the complement of $\mathbb{R}^+ \cup \{0\}$ in $\mathbb{R}$ the **negative numbers** and denote it by $\mathbb{R}^-$.   □

Here are some easy exercises to help you familiarize yourself with the order axiom.

8.3.1   EXERCISE

For all real numbers a, one and only one of the following holds:

- $a \in \mathbb{R}^+$.
- $a = 0$.
- $a \in \mathbb{R}^-$.   □

8.3.2   EXERCISE

The standard "sign rules" for addition and multiplication hold. The fact that two positive numbers added or multiplied together yield a positive number is the first provision of the order axiom. Here are some other "sign rules." Let a and b be real numbers. Prove the following.

1. If $a, b \in \mathbb{R}^-$, then $a + b \in \mathbb{R}^-$.
2. If $a, b \in \mathbb{R}^-$, then $a \cdot b \in \mathbb{R}^+$.
3. If $a \in \mathbb{R}^+$ and $b \in \mathbb{R}^-$, then $a \cdot b \in \mathbb{R}^-$.

(*Hint:* Keep in mind Theorem 8.2.5 as well as the provisions of the order axiom.)   □

8.3.3   EXERCISE

Prove the following facts.

1. $1 \in \mathbb{R}^+$ and -1 in $\mathbb{R}^-$.
2. If $a \in \mathbb{R}^+$, then $\frac{1}{a} \in \mathbb{R}^+$, and if $a \in \mathbb{R}^-$, then $\frac{1}{a} \in \mathbb{R}^-$. □

We said in Section 8.2 that once we have the concept of the positive numbers we can define an order on $\mathbb{R}$.

8.3.4   DEFINITION

We define a relation $\leq$ on $\mathbb{R}$ as follows. For all real numbers x and y, we say that $x \leq y$ if $y - x \in \mathbb{R}^+ \cup \{0\}$.

Here are some important theorems about the order relation on $\mathbb{R}$. They are followed by some theorems about the way that $\leq$ and the arithmetic operations interact.

8.3.5   THEOREM (Total ordering)

Show that $\leq$ is a total order on $\mathbb{R}$. □

8.3.6   COROLLARY (Trichotomy)

For all real numbers a and b, one and only one of the following is true.

• $a < b$
• $a = b$
• $a > b$ □

8.3.7   THEOREM (Order and arithmetic)

Let a, b, c, and $d \in \mathbb{R}$.

1. If $a > b$ and $c \geq d$, then $a + c > b + d$.
2. If $a > b$ and $c > 0$, then $ac > bc$.
3. If $a > b > 0$, and $c \geq d > 0$, then $ac > bd$. □

You can now derive all sorts of arithmetic facts. Here are just a few for you to try your hand at.

8.3.8   PROBLEM

Prove the following arithmetic facts.

1. Let x be a real number. If $x > 1$, then $x^2 > x$.
2. Let x be a positive real number. If $x < 1$, then $x^2 < 1$.
3. Let a and b be nonzero real numbers that are either both positive or both negative. If $a \leq b$ then $\frac{1}{a} \geq \frac{1}{b}$. (What happens if a is negative and b is positive?) □

Next, we introduce the absolute value function and prove some of its basic properties.

8.3.9   DEFINITION

Let a be a real number. We define the **absolute value** of a (denoted by $|a|$) as follows:

$$|a| = \begin{cases} a & \text{if } a \in \mathbb{R}^+ \cup \{0\}, \\ -a & \text{if } a \in \mathbb{R}^-. \end{cases}$$

Establish the following facts about $|\cdot|$.

8.3.10   THEOREM

Let a and b be real numbers.

1. $|a| = \max(a, -a)$. (See Problem 5.6.11 for a definition.)
2. $|a| \geq 0$.
3. $|a \cdot b| = |a||b|$.
4. $|a| = |-a|$.
5. $|a + b| \leq |a| + |b|$.
6. $|a - b| \geq ||a| - |b||$.
7. Let ϵ be any positive number. Then $|a - b| < \epsilon$ if and only if $a \in (b - \epsilon, b + \epsilon)$.

$\square$

In the order axiom we assume that for all positive real numbers a, a and $-a$ are in disjoint subsets of $\mathbb{R}$ and thus that they are different. This takes care of statement III. We have also addressed the question of positivity, which puts us a little farther on the road to discussing square roots. What about the infinitude of the set of real numbers?

We now need to appeal to the natural numbers. The natural numbers can be constructed using only set theory (see Appendix A), and since we are accepting axiomatic set theory as part of our axiomatic structure, theorems about natural numbers (such as counting and induction results) are available for our use; however, there is a subtlety here. The set-theoretic "natural numbers" are not necessarily connected with the set $\mathbb{R}$ that we are studying in this chapter. We cannot *assume* that one is a subset of the other. However, using the natural numbers constructed from set theory, we will now identify a subset of $\mathbb{R}$ that "looks just like" them. (For a more sophisticated look at this construction see Appendix B.)

Let a be a positive real number. Let $k \in \mathbb{N}$. We define

$$ka = \underbrace{a + a + a + \cdots + a}_{k \text{ times}}.$$

Note that ka is **not** assumed to be k multiplied by a. In fact, it would not make any sense to interpret it this way, since k is not even an element of $\mathbb{R}$. k serves only as a counter; we are adding a to itself this number of times. Be sure to keep this in mind as you prove the next lemma.

8.3.11   LEMMA

Let a be a positive real number. Then for all distinct pairs of natural numbers n and m

$$na \neq ma.$$

(*Hint*: Assume that n > m and that na = ma. Deduce that $(n - m)a = 0$ and derive a contradiction. You will need both provisions of the order axiom.)   □

8.3.12   THEOREM

Show that $\mathbb{R}$ contains a sequence of distinct terms and is therefore infinite.   □

If we take a to be 1 in Lemma 8.3.11, the sequence that we obtain

$$1, 1 + 1, 1 + 1 + 1, 1 + 1 + 1 + 1, \ldots, k1, \ldots$$

is mathematically identical to the set-theoretic natural numbers we spoke of above. That is, it is isomorphic to the natural numbers. (See Appendix A for a set of axioms of the natural numbers.) From now on we can think of the natural numbers as sitting inside $\mathbb{R}$ and list them in the usual way: $\{1, 2, 3, \ldots\}$. We obtain the integers by including 0 and the additive inverses of the natural numbers. Finally, we get the rational numbers by taking quotients of integers. Real numbers that are not rational, of course, we call irrational. We have shown that $\mathbb{R}$ contains the natural numbers. The field axioms tell us that we can then construct integers and rational numbers. How do we know there *is* anything else? On the face of it, we do not. We have to do further work to show that there are irrational numbers.

8.3.13   PROBLEM

1. Show that there is no rational number whose square is 2. (*Hint*: Proceed by contradiction. Remember: A rational number is a number that can be written as a ratio of integers. One classic proof assumes $\sqrt{2}$ is rational, that is, $\sqrt{2} = \frac{m}{n}$. What can you say about the prime factors of m^2 and $2n^2$ and what does this tell you?)

2. This result does not in itself prove for us that there are irrational numbers. Explain why not.   □

Before we introduced the order axiom, we saw a mathematical structure (very different from the real numbers) that satisfied the field axioms, and we were thus able to conclude that we could not prove several important properties of the real numbers by using only the field axioms. Some of those problems have been resolved by accepting the order axiom. As you will see in the next problem, the rationals (considered not as a subset of $\mathbb{R}$, but as a mathematical structure in themselves) satisfy all the axioms given so far. This tells us that we will need a further axiom to distinguish the real numbers from the rational numbers.

8.3.14   PROBLEM

Consider the rational numbers under ordinary addition and multiplication of real numbers.

1. Verify that the rational numbers under addition and multiplication satisfy the field axioms.

2. Verify that the rational numbers satisfy the order axioms.

3. Assuming only the field axioms and the order axiom, will you be able to prove that every positive real number has a square root? Explain.   □

8.4   The Least Upper Bound Axiom

We noted in the section on the order axiom that the rational numbers satisfy the field axioms and the order axiom. However, since $\sqrt{2}$ is not a rational number, it is not true that every positive rational number has a (rational) square root. Thus we cannot hope to prove that every positive real number has a square root using only the field axioms and the order axiom. We need one more axiom. We need an axiom that will "fill the holes" left by the rational numbers. The axiom that we need is called the least upper bound axiom. First, we present an exercise that will help you review some of the concepts introduced in Section 4.2.

8.4.1   EXERCISE

Consider the following subsets of $\mathbb{R}$. Which of them has a least upper bound? A greatest element? A greatest lower bound? A least element?

1. $\mathbb{N}$ 2. $[0, 1]$ 3. $(-2, 3)$
4. $(-\infty, 7)$ 5. $\{1, \frac{1}{2}, \frac{1}{3}, \frac{1}{4}, \ldots\}$ 6. $\emptyset$

We can now state the final axiom for the real numbers.

AXIOM III (Least upper bound axiom)

$\mathbb{R}$ has the least upper bound property. That is, every nonempty subset of $\mathbb{R}$ that is bounded above has a least upper bound.

Having proved Theorem 4.2.26, we have the following corollary.   □

8.4.2   COROLLARY

Every nonempty subset of $\mathbb{R}$ that is bounded below has a greatest lower bound.   □

8.4.3   THEOREM

Let B be a nonempty subset of $\mathbb{R}$ that is bounded above. Let $b \in \mathbb{R}$ be an upper bound for B. The following statements about b are equivalent.

    i. b is the least upper bound of B.

    ii. For each positive number ϵ there exists $x \in B$ such that $|x - b| < \epsilon$.

    iii. For each positive number ϵ, there exists $x \in B$ such that $x \in (b - \epsilon, b + \epsilon]$. ☐

Here are some very useful consequences of the least upper bound property of $\mathbb{R}$. The first says that there are arbitrarily large natural numbers. The second says that there are arbitrarily small multiplicative inverses of natural numbers.

8.4.4 THEOREM (The Archimedean property of $\mathbb{R}$)

For every real number x there is an integer n such that $n > x$.

    (*Hint:* Proceed by contradiction. Conclude that $\mathbb{N}$ must be bounded above and must therefore have a least upper bound. Use the least upper bound axiom and Theorem 8.4.3 to derive a contradiction.) ☐

8.4.5 THEOREM

For every positive number ϵ there exists a positive integer n such that $1/n < \epsilon$. ☐

We are finally in a position to prove that every positive real number has a square root.

8.4.6 THEOREM

For every positive real number a there exists a real number x such that $x^2 = a$. ☐

    Hints for Proving Theorem 8.4.6: This is an existence proof. Consider the set

$$S = \{y \in \mathbb{R} : y^2 < a\}$$

Show that S is nonempty and bounded above. Conclude that it has a least upper bound, x. To prove that $x^2 = a$, show that $x^2 \not< a$ and $x^2 \not> a$, then appeal to trichotomy.

8.4.7 THEOREM

There exists an irrational number. ☐

We now have a set of axioms that gives us properties I–V that we stated in Section 8.2. In fact, we have all the axioms we need to completely describe the real number system, though this is not obvious. We conclude this chapter by looking at how integers, rational numbers, and irrational numbers fit together in $\mathbb{R}$.

8.4.8 THEOREM

For any $x \in \mathbb{R}$ there is an integer n such that $n \le x < n + 1$.

    (*Hint:* You may find Theorem 6.1.4(2) useful.) ☐

In other words, every real number is between two *consecutive* integers.

8.4.9   LEMMA

For any $x \in \mathbb{R}$ and any positive integer N there exists an integer n such that

$$\frac{n}{N} \le x < \frac{n+1}{N}. \qquad \square$$

These facts imply the following two (very useful) results: that every real number can be approximated as closely as desired by a rational number and that between any two distinct real numbers there is a rational number.

8.4.10   THEOREM

If $x \in \mathbb{R}$ and $\epsilon > 0$, there exists a rational number r such that $|x - r| < \epsilon$. $\qquad \square$

8.4.11   THEOREM

Let x and y be real numbers with $x < y$. Then there exists a rational number r satisfying $x < r < y$.

Proof Sketch.   Provide the details necessary to support the following steps.

Step I. Choose a positive integer q so that $\frac{1}{q} < x - y$.

Step II. Choose an integer p so that $\frac{p}{q} \le y < \frac{p+1}{q}$.

Step III. Now show that $x < \frac{p}{q} < y$. $\qquad \square$

There is also a complementary fact about irrational numbers.

8.4.12   THEOREM

Let x and y be real numbers with $x < y$. Then there exists an irrational number s satisfying $x < s < y$.

(*Hint:* The quickest way to get this is to use a cardinality argument. But it is perhaps most instructive to do it "constructively." Start with $\sqrt{2}$. Make it "small" by dividing by a sufficiently large natural number. Show that this small number k is irrational. Then show that $x + k$ is irrational.) $\qquad \square$

These last two theorems tell us that any open interval in $\mathbb{R}$ contains both rational and irrational numbers. It is somewhat amazing to contemplate how much we have gotten from so few assumptions. The set we started with was not even assumed to be infinite. We have now shown that it contains all the familiar characters: the natural numbers, the integers, the rational numbers and the irrational numbers—and that these characters behave in familiar ways.

It is certain that none of the results that you have proved so far in this chapter were surprising or unfamiliar to you. That's good! That means that the axioms are doing what they are supposed to be doing—describing the real numbers. There are surprises to be had in the real numbers, but they are not easy consequences of the axioms. (The easy consequences we all know about!) You will have to go further in your study of mathematics to get into unfamiliar territory. However, we now have some evidence that the axioms are good ones. When you later learn something surprising from them, you can be confident in that result.

8.5   Sequence Convergence in $\mathbb{R}$

We have now considered in detail the axioms for the real numbers and some of their immediate consequences. It would be unsatisfying to end the book without tying them in with some of the other themes that we have explored. We could proceed in a number of directions; I choose to consider the convergence of sequences of real numbers. Though a great deal can be said about this topic, our goals will be modest. We will start by using our intuition about the word "convergence" to help us define what is meant by the convergence of a sequence, and we will end by proving that $\mathbb{R}$ has an important property called *completeness*.

The sentence

> *Fifteen thousand spectators* **converged** *on the coliseum to watch the state championships*

brings to mind a picture of crowds of people coming toward a sports stadium from every direction. The word *converge* gives the image of things moving toward a fixed spot. With this image in mind, we think of a sequence as converging if its terms approach a particular point more and more closely as they appear farther and farther out in the sequence.

Consider the following example.

$$1, \frac{1}{2}, \frac{2}{3}, \frac{3}{4}, \frac{4}{5}, \cdots \frac{n}{n+1} \cdots$$

The terms that appear far out in the sequence (for instance, $\frac{9,999,999}{10,000,000}$) are very close to 1. We wish to say that this sequence converges to 1.

Here are two more examples:

$$3, -3, 3, -3, 3, -3, \ldots$$

$$1, 2, 3, 4, 5, \ldots$$

Our intuition tells us that these should not fit our definition of convergence. A sequence that oscillates back and forth between two values is not approaching any one number. The sequence of natural numbers is marching off to infinity and its terms are not approaching any one number, either.

Our intuitive description of a convergent sequence is, unfortunately, not very useful from a mathematical point of view. What do we mean when we say "approach . . . more and more closely"? For starters, we are talking about distance. In the real numbers, distance is measured using absolute values.

8.5.1 DEFINITION

Let x and y be real numbers. We define the **distance** between x and y to be $|x - y|$.

Here are some important properties of distance in $\mathbb{R}$.

8.5.2 THEOREM

Let a, b, and c be real numbers.

1. The distance from a to b is always positive, unless $a = b$ in which case it is zero.
2. The distance from a to b is the same as the distance from b to a.
3. The distance from a to b plus the distance from b to c is greater than or equal to the distance from a to c. (This last property is called the **triangle inequality**.) ☐

We can now say what it means for a sequence to converge.

8.5.3 DEFINITION

Let (a_n) be a sequence of real numbers. We say that (a_n) **converges** to a number L provided that for all real numbers $\epsilon > 0$ there exists $N \in \mathbb{N}$ so that for all $n > N$

$$|a_n - L| < \epsilon.$$

We denote the convergence of (a_n) to L by $a_n \to L$ or $\lim_{n \to \infty} a_n = L$.

8.5.4 EXERCISE

State in positive terms what it means to say that (a_n) *does not* converge to L. (That is, negate Definition 8.5.3.) ☐

8.5.5 EXERCISE

1. Show that the sequence $-1, -1, -1, -1, \ldots$ converges.
2. Show that the sequence $1, -\frac{1}{2}, \frac{1}{3}, -\frac{1}{4}, \frac{1}{5}, -\frac{1}{6}, \ldots$ converges.
3. Show that the sequence $10, \frac{1}{2}, 10, \frac{1}{3}, 10, \frac{1}{4}, \ldots$ does not converge. ☐

Prove the following standard (and useful) facts about convergence of sequences of real numbers.

8.5.6 THEOREM

Prove that every convergent sequence of real numbers is bounded. ☐

8.5.7 THEOREM

Let (a_i) and (b_i) be sequences of real numbers. Suppose that (a_i) converges to a and (b_i) converges to b. Let k be any real number. Then

1. $(a_i + b_i)$ converges to $a + b$.

2. (ka_i) converges to ka.

3. Presuming that a_i is never zero and $a \neq 0$, prove that $\frac{1}{a_i}$ converges to $\frac{1}{a}$. (*Hint:* Use the fact that (a_i) is a bounded sequence.)

4. $(a_i b_i)$ converges to ab. (*Hint:* Express $ab - a_i b_i$ as $ab - ab_i + ab_i + a_i b_i$ and remember that (b_i) is a bounded sequence. Then work at it a bit.)

5. Presuming all quotients make sense (b_i is never zero and b is not zero) $\frac{a_i}{b_i}$ converges to $\frac{a}{b}$. (*Hint:* Don't work too hard here. Rest on your laurels!)   □

8.5.8   THEOREM

Let (a_i) be a bounded, monotonic sequence of real numbers. Show that (a_i) is convergent.

(*Hint:* Draw a picture. Guess what the limit of the sequence should be, then prove your conjecture.)   □

As a consequence of this theorem and Theorem 5.5.25 we have the following theorem and corollary.

8.5.9   THEOREM

Let (a_i) be any bounded sequence of real numbers. Show that (a_i) has a convergent subsequence.   □

8.5.10   COROLLARY

Every bounded infinite subset of $\mathbb{R}$ has a convergent sequence of distinct terms.   □

We know how to measure distances in $\mathbb{R}$. We can measure distances in other mathematical structures, as well: For instance, we define the distance between two points in $\mathbb{R}^2$ and $\mathbb{R}^3$ to be the length of the straight-line segment joining them. As a result of our interest in measuring distances, we define a mathematical structure based on the concept of distance.

8.5.11   DEFINITION

Let M be a nonempty set. Let $d : M \times M \to \mathbb{R}$ be a function satisfying the following properties: For all x, y, and z in M,

- $d(x, y) \geq 0$. $d(x, y) = 0$ iff $x = y$.
- $d(x, y) = d(y, x)$.
- $d(x, z) \leq d(x, y) + d(y, z)$.

M together with d is called a **metric space**. The function d is called the **metric** or **distance function** on M.

You have already shown that $\mathbb{R}$ together with the distance function $d(x, y) = |x - y|$ is a metric space (Theorem 8.5.2); there are many other examples.

8.5.12 EXERCISE

Let M be any nonempty set, and define

$$d(x, y) = \begin{cases} 0 & \text{if } x = y, \\ 1 & \text{if } x \neq y, \end{cases}$$

for all $x, y \in M$. Show that M together with d is a metric space.   □

8.5.13 EXAMPLE

As I said above, the usual straight-line distance between points in $\mathbb{R}^2$ or $\mathbb{R}^3$ is also a metric. That is, given two points (x_1, y_1) and (x_2, y_2) in $\mathbb{R}^2$, the function $d : \mathbb{R}^2 \rightarrow \mathbb{R}$ given by

$$d((x_1, y_1), (x_2, y_2)) = \sqrt{(x_1 - x_2)^2 + (y_1 - y_2)^2}$$

is a distance function. Likewise, $d : \mathbb{R}^3 \rightarrow \mathbb{R}$ given by

$$d((x_1, y_1, z_1), (x_2, y_2, z_2)) = \sqrt{(x_1 - x_2)^2 + (y_1 - y_2)^2 + (z_1 - z_2)^2}$$

is a distance function.

All the necessary properties are easy to prove with the exception of the triangle inequality, which is challenging. (See the Questions to Ponder.)   ■

There can be different distance functions on the same set.

8.5.14 EXERCISE

Consider the following function on $\mathbb{R}^2 \times \mathbb{R}^2$:

$$d((x_1, y_1), (x_2, y_2)) = |x_1 - x_2| + |y_1 - y_2|.$$

1. Show that d is a metric on $\mathbb{R}^2$.

2. This is sometimes called the "taxicab metric." To see why, give a geometric interpretation of the distance between two points in the plane.   □

Sequence convergence can be defined in an arbitrary metric space.

8.5.15 DEFINITION

Let (x_i) be a sequence in a metric space. (x_i) is said to **converge to** x if for every real number $\epsilon > 0$ there exists $N \in \mathbb{N}$ such that if $n > N$, then $d(x_n, x) < \epsilon$.

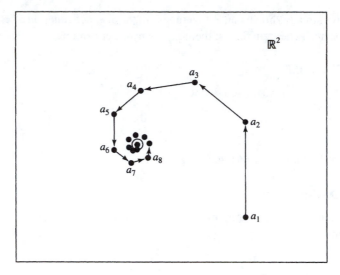

Figure 8.1   Sequence convergence in $\mathbb{R}^2$

8.5.16   EXERCISE

Verify that the definition of convergence we gave for sequences in $\mathbb{R}$ coincides with this more general definition when d is the usual distance function $d(x, y) = |x - y|$.   □

8.5.17   THEOREM

Let M be a metric space with metric d. Let (a_i) be a sequence in M. Then (a_i) converges to a if and only if every subsequence of (a_i) converges to a.   □

Our definition of convergence says that a sequence converges if its terms get closer and closer to a given point. This, of course, implies that the terms get closer and closer together. A sequence in which terms get closer and closer together is said to be a Cauchy sequence.

8.5.18   DEFINITION

Let (a_i) be a sequence in a metric space. We say that (a_i) is **Cauchy** if for every real number $\epsilon > 0$ there exists $N \in \mathbb{N}$ such that for all natural numbers n and m with $n > N$ and $m > N$

$$d(a_n, a_m) < \epsilon.$$

8.5.19   THEOREM

Let M be a metric space with distance function d. Let (a_i) be a sequence in M converging to a. Show that (a_i) is Cauchy.

(*Hint:* You will need to use the triangle inequality.)   □

8.5.20   DEFINITION

A metric space is said to be **complete** if every sequence that is Cauchy is also convergent.

Not all metric spaces are complete.

8.5.21   EXERCISE

The rational numbers by themselves form a metric space under the usual distance function. Show that this metric space is not complete by constructing a Cauchy sequence that does not converge to any rational number.

(*Hint*: Construct a sequence of rational numbers that "converges to" an irrational number such as $\sqrt{2}$.)   □

The rational numbers are not complete because they are "full of pinholes." A Cauchy sequence can therefore be constructed that approaches one of the pinholes. The real numbers, on the other hand, *have no holes*—thus every sequence that gets closer and closer together actually converges to some real number. The following results will allow us to show that $\mathbb{R}$ is a complete metric space.

8.5.22   LEMMA

Show that every Cauchy sequence in $\mathbb{R}$ (under the usual metric) is a bounded sequence.

□

8.5.23   LEMMA

Let M be a metric space with distance function d. Let (a_i) be a Cauchy sequence in M. Suppose that (a_i) has a subsequence converging to a. Then (a_i) converges to a.   □

8.5.24   THEOREM

Show that $\mathbb{R}$ is complete under the usual metric.   □

■ PROBLEMS

1. Let a and b be real numbers. Prove the following algebraic identities.

   (a) $(a + b)^2 = a^2 + 2ab + b^2$

   (b) $(a - b)^2 = a^2 - 2ab + b^2$

2. In Exercise 8.3.2 you proved several "sign rules" for adding and multiplying real numbers. None of those allowed for the possibility that one or both of the numbers could be 0. Formulate and prove all such possible "sign rules." There are eight of them, but each should be an easy consequence of the previously proved rules and the field axioms.

3. Let a, b, c, and $d \in \mathbb{R}$. Prove the following.

   (a) If $a \geq b$ and $c \geq d$, then $a + c \geq b + d$.

   (b) If $a \geq b > 0$, and $c \geq d > 0$, then $ac \geq bd$.

■ QUESTIONS TO PONDER

1. Try to prove that $\mathbb{R}^2$ under the usual straight-line distance is a metric space. That is, prove that the function $d : \mathbb{R}^2 \to \mathbb{R}$ given by

$$d((x_1, y_1), (x_2, y_2)) = \sqrt{(x_1 - x_2)^2 + (y_1 - y_2)^2}$$

   is a distance function.

2. We have spoken in various contexts about the preservation of mathematical structure through functions. Suppose we have two metric spaces, (X, d) and (Y, k), and a function $f : X \to Y$. Suppose that given any convergent sequence in X, its image converges in Y. Symbolically, if $a_n \to a$ in X, then:

$$a_n \to a \quad \text{implies that} \quad f(a_n) \to f(a).$$

   What does this tell us about the function f? For instance, think of the case in which both X and Y are $\mathbb{R}$ with the usual distance. What can you say about the graph of f? Draw pictures. Make conjectures. Prove theorems.

Axiomatic Set Theory

> I decided that I was really a physicist instead of a mathematician when I discovered that all of mathematics is composed entirely of brackets—take the brackets away, and there's nothing left.—*A theoretical physicist*

This appendix is designed to give an outline of the axiomatic basis for the theory of sets. The purpose of this is to put our intuitive ideas about sets on a rigorous footing while avoiding logical difficulties such as Russell's paradox. The emphasis of the system of axioms is therefore on the *existence* of various types of sets. Set theory only deals with sets whose existence can be explicitly demonstrated on the basis of the axioms.

Our treatment will be reasonably rigorous, but incomplete. Readers who are interested in pursuing axiomatic set theory further should consult a full text on the subject.

Remember as you work through this appendix that we cannot employ the familiar machinery of sets and their operations until we build that machinery out of the axioms.

A.1 Elementary Axioms

Like any axiomatic system, set theory has undefined terms, *set* and *element,* together with the symbol $\in$. The fundamental relationship between objects in the theory is membership: $x \in y$, meaning that x is an element of the set y. From a formal point of view, the only meaning we ascribe to these concepts is the meaning given to them by the axioms. The axioms will specify how the undefined terms behave and relate to one another. We are now ready to state our first two axioms.

AXIOM I (Axiom of existence)

There exists a set. ☐

197

AXIOM II (Axiom of extensionality)

Two sets are equal if and only if they have exactly the same elements.   □

The first axiom tells us that the subject of set theory is not an empty one! Note that the idea of "existence" here is purely mathematical and may or may not have anything to do with physical existence. The second axiom tells us that a set is completely characterized by its elements: Two sets A and B are equal provided that, for every x, $x \in A$ if and only if $x \in B$. In other words, sets have no "additional structure" beyond their membership.

These first two axioms are only a start. By themselves, they cannot demonstrate the existence of any *particular* set. We remedy this in an interesting way by means of a third axiom.

AXIOM III (Axiom of specification)

If A is a set and $Q(x)$ is some predicate, then there exists a set B consisting of exactly those elements $x \in A$ for which $Q(x)$ is true.   □

This is one of the most useful and far-reaching of the set axioms. It tells us that, once we have a set, we can form subsets according to any rule we please. For example, once we have the natural numbers, we will be able to use the axiom of specification to form the set of even natural numbers, the set of prime natural numbers, and so on.

Strictly speaking, the axiom of specification is not a single statement about sets, but rather an infinite family of statements, one for each specific predicate $Q(x)$. It is thus sometimes called an **axiom schema**, a rule for generating an unlimited number of axioms of a specific form. We will ignore this logical nicety and think of the axiom of specification in the intuitive way, as a single fact about sets.

We had better formalize our use of the term "subset."

A.1.1 DEFINITION

A set B is a **subset** of a set A if every element of B is also an element of A. We write this as $B \subseteq A$.

A.1.2 EXERCISE

Show that every set is a subset of itself.   □

Armed with specification, we can now get our hands on an actual set.

A.1.3 THEOREM

There exists a unique set with no elements. (*Hint:* Take an existing set and form the subset of those elements x for which $x \neq x$. Your proof should explicitly use all three of the axioms given so far.)   □

A.1.4   DEFINITION

The unique set with no elements is called the **empty set** (or **null set**) and is denoted by $\emptyset$.[1]

Specification is a powerful tool for building new sets out of old ones. We will use it frequently as we define various familiar items from the set-theoretic inventory. Every newly defined set must be accompanied by a proof of its existence based on the axioms, and this proof frequently makes use of specification. (Uniqueness, when required, is usually proved by means of the axiom of extensionality.)

A.1.5   EXERCISE

Let A and B be sets for which $B \subseteq A$. Define the **complement** of B in A and use the axioms to show that it exists and is unique.   □

A.1.6   DEFINITION

Let A and B be sets. The **intersection** $A \cap B$ is the subset of A consisting of those elements $x \in A$ for which $x \in B$.

A.1.7   EXERCISE

Show that $A \cap B$ exists and that $A \cap B = B \cap A$. (This last fact is not completely trivial, since $A \cap B$ is defined as a subset of A, while $B \cap A$ is a subset of B.)   □

A.1.8   EXERCISE

Let A be a set. Show that $A \cap A = A$ and $A \cap \emptyset = \emptyset$.   □

Notice that we have only defined intersection for *pairs* of sets. We will postpone a general definition until we have more axioms.

The three axioms given so far are sufficient for Russell's paradox.

A.1.9   THEOREM

There does not exist a set containing all sets. (*Hint:* See Section 2.6 for a more complete discussion. Proceed by contradiction.)   □

We cannot simply define a set by a predicate (e.g., "the set of everything red"). Instead, we first need a larger set ("the set of socks") whose existence is already guaranteed in some other way; then we can use the predicate to get at a subset ("the set of red socks") by means of specification. This allows us to use predicates to define sets in a safe and limited fashion, avoiding Russell's paradox.

---

[1] Or, more picturesquely, by { }.

On the other hand, the axioms so far cannot yet guarantee the existence of any set except the empty set! We need more axioms for a useful theory.

Suppose we have a few things—in our case, sets—that are already known to "exist" mathematically. We would like to be able to turn that collection into an axiomatically justifiable set, to build the set out of its elements. It will suffice to be able to do this for just two elements, which are themselves sets.

AXIOM IV (Axiom of pairing)

If A and B are sets, then there is a set C such that $A \in C$ and $B \in C$. □

Note that the axiom does not say that C contains *only* A and B, but we can take care of that detail by using specification.

A.1.10 THEOREM

If A and B are sets, then there exists a set, denoted $\{A, B\}$, that contains only A and B as elements. □

In general, we will understand $\{A, B, C, \ldots\}$ to mean a set (whose existence must be demonstrated) containing exactly $A, B, C, \ldots$ as elements.

A.1.11 THEOREM

For any set A, the set $\{A\}$ exists. □

A.1.12 QUESTION

Is $\emptyset$ the same as $\{\emptyset\}$? Why or why not?

A.1.13 EXERCISE

Demonstrate, using the axioms, the existence of at least six distinct sets. □

A note on style: In the axiom of existence, we assumed the existence of some unspecified set; but the only place we actually used the axiom was to establish the existence of $\emptyset$ by specification. We might instead have assumed the existence of $\emptyset$ directly. In a similar way, the axiom of pairing merely assumes the existence of some set C containing both A and B, and then we obtained the specific set $\{A, B\}$ by specification. I could have used existence of $\{A, B\}$ as a more specific version of the axiom. I have chosen so far to make the statements of our axioms as "weak" as possible, proving stronger and more specific statements as theorems. Other books make other choices.

A.1.14 EXERCISE

Suppose A, B, and C are all sets. Can you demonstrate the existence of the set $\{A, B, C\}$? □

This exercise suggests that we have not yet finished with our axioms! We might have tried to show the existence of $\{A, B, C\}$ in this way: We can certainly make the sets

$\{A, B\}$ and $\{B, C\}$ by the axiom of pairing. Is there any way to "join" these sets together? Applying the axiom of pairing again yields $\{\{A, B\}, \{B, C\}\}$, which is not quite good enough. (Why not?) Intuitively, we need to take the *union* of $\{A, B\}$ and $\{B, C\}$, and we will require another axiom to do this.

In fact, we want to be able to take the union of a whole bunch of sets, not just two at a time. But we had better be careful about this and specify what we mean by a "bunch" of sets. For example, if we are allowed to form a set by taking the union of *all* sets, we could then construct the set of all sets—which we know does not exist. (Think why this is so.) Therefore, an axiom that permitted us to form such a union would yield an *inconsistent* axiomatic system, one in which a logical contradiction can be derived from the axioms.

The safe thing to do is to allow ourselves to take the union of any *set* of sets.

AXIOM V (Axiom of union)

Suppose A is a set whose elements are all sets. Then there exists a set $\cup A$ consisting of all x such that $x \in B$ for some $B \in A$. □

We call this set $\cup A$ the **union** over the set A. Once we have the axiom of union, we can prove lots of things.

In this instance, we are phrasing the axiom in the "strong and specific" version. Think of a weaker axiom from which we could prove our axiom of union as a theorem, perhaps using specification.

A.1.15 THEOREM

Suppose A and B are sets. Then there exists a unique set $A \cup B$ consisting of all x such that $x \in A$ or $x \in B$. □

A.1.16 QUESTION

Can you explain why a new axiom is required to define $A \cup B$, but not $A \cap B$?

A.1.17 EXERCISE

Resolve Exercise A.1.14 by showing the existence of $\{A, B, C\}$, given that A, B, and C are all sets. □

A.1.18 EXERCISE

Construct, on the basis of the axioms, a set containing exactly three elements. (How is this different from the previous problem?) □

When we say "construct a set," we mean to specify the set in such a way that its existence is guaranteed by the axioms. Thus, "the set of all primary colors" is not a correct solution to the previous exercise, since its existence cannot be justified on the basis of our axiomatic system.

A.1.19 EXERCISE

Construct a set containing exactly seven elements. □

By now you have probably noticed something curious about the sets that arise in axiomatic set theory. Every nonempty set that you have constructed has as its elements other sets. In fact, *axiomatic set theory does not assume the existence of set elements that are not themselves sets.* The theory does not deny that "nonset" elements (sometimes called "atoms") might exist, but it does not affirm their existence either. We can get along without them. All of the sets required to do ordinary mathematics—including everything in this book—are built out of the empty set using various axioms. (See the quotation at the head of this appendix.)

Here is the promised general definition of intersection.

A.1.20   DEFINITION

Suppose A is a set whose elements are all sets. The set $\cap A$, called the **intersection** over the set A, is defined as the set of all x such that $x \in B$ for every $B \in A$.

A.1.21   THEOREM

For any set A whose elements are sets, $\cap A$ exists.   □

A.1.22   EXERCISE

Show that $A \cap B$, defined earlier, is just $\cap\{A, B\}$.   □

Once we have defined union and intersection, we can prove various properties of these, as was done in Section 2.3.

The axioms of existence and extensionality get us started. The axiom of specification allows us to form subsets at will—that is, to use established sets to form smaller ones. The axioms of pairing and union allow us to build up larger sets from those at hand.

Another way of making a new and larger set out of a set A is to take the collection of all subsets of A. This is called the **power set** of A and is denoted $\mathcal{P}(A)$. The existence of such a set cannot in general be assured by the axioms so far, so we introduce a new axiom.

AXIOM VI (Axiom of the power set)

For any set A, there exists a set $\mathcal{P}(A)$ such that $B \in \mathcal{P}(A)$ if and only if $B \subseteq A$.   □

This axiom is frequently used to build some useful mathematical structure. Here is an instance of this sort of construction.

A.1.23   DEFINITION

For any a and b, define the **ordered pair** $(a, b) = \{\{a\}, \{a, b\}\}$.

A.1.24   EXERCISE

Show that $(a, b) = (c, d)$ if and only if $a = c$ and $b = d$. This is the essential property of ordered pairs.   □

Technically speaking, we can only ensure the existence of such an object if a and b are themselves sets. This will be good enough.

A.1.25   DEFINITION

Suppose A and B are sets. The **Cartesian product** $A \times B$ is the set of all ordered pairs (a, b) such that $a \in A$ and $b \in B$.

A.1.26   THEOREM

For any two sets A and B, $A \times B$ exists. (*Hint:* Form $\mathcal{P}(\mathcal{P}(A \cup B))$, which will contain all of the ordered pairs together with lots of other things. Get rid of the other things using specification.) □

Relations and functions are simply subsets of one Cartesian product or another. This theorem can therefore be seen as the axiomatic foundation for Chapters 4 and 5.

A.2   The Axiom of Infinity

In this section we will give a set-theoretic construction of the set $\mathbb{N}$ of natural numbers. Each natural number will, of course, be defined as a set. Our construction will require a new axiom.

It is easy to see why. The axioms given so far are sufficient to show the existence of *finite* sets of arbitrary size. See the following example.

A.2.1   EXAMPLE

On the basis of the axioms given so far, construct a set with more than one million elements. ∎

The difficulty is that we cannot yet prove the existence of any *infinite* set. Suppose we begin to build sets using the axioms we have given so far. We begin with the empty set and then apply our axioms in various combinations to build further sets. At any given stage in this process we can only have built a finite assortment of sets that are each finite, and each of the axiomatic operations (specification, pairing, union, or power set) on this assortment can yield only a finite set. Because $\mathbb{N}$ is infinite, we can never reach it in this way.

Put somewhat differently, we can construct sets that are as large as *any given* natural number n, but we cannot construct a set large enough to contain *all* natural numbers.

If you have worked through Chapter 3, you will recall that a similar difficulty arises in the development of mathematical induction. To resolve this, it is necessary to adopt the Induction Axiom, which explicitly allows us to deduce infinitely many things in a finite number of logical steps. The new axiom we will adopt here will be similar in some respects to the Induction Axiom. Indeed, we will be able to use it to prove the Induction Axiom, thereby giving a set-theoretic foundation for Chapter 3.

First, we will need a couple of definitions.

A.2.2   DEFINITION

For any set A, the **successor** of A, denoted $S(A)$, is defined to be $S(A) = A \cup \{A\}$. (You should, of course, figure out why such a set must exist.)

A.2.3   EXERCISE

Find $S(\emptyset)$, $S(S(\emptyset))$, $S(S(S(\emptyset)))$, and $S(S(S(S(\emptyset))))$. How many distinct elements are in each set?   □

To define the natural numbers $\mathbb{N}$ within axiomatic set theory, we will have to come up with objects to play the roles of the various numbers 0, 1, 2, 3, and so on. These objects will be sets. The identification of number and set will be arbitrary, but things work out most easily if, for example, the set that plays the role of 7 is a set with seven elements. The exercise above suggests a way to do this:

$$0 \leftrightarrow \emptyset,$$
$$1 \leftrightarrow S(\emptyset),$$
$$2 \leftrightarrow S(S(\emptyset)),$$
$$3 \leftrightarrow S(S(S(\emptyset))),$$

and so on. This way of identifying sets with numbers is especially desirable because it makes defining other properties of $\mathbb{N}$ especially easy later on.

In axiomatic set theory, it is common practice to include 0 in $\mathbb{N}$, although 0 is not considered to be a natural number in most other mathematical contexts. I will follow this peculiarly inconsistent convention, including 0 in $\mathbb{N}$ throughout the appendices but excluding 0 from $\mathbb{N}$ in the main text of the book.

A.2.4   EXERCISE

Explicitly write down the set 5.   □

A.2.5   EXERCISE

Show that $7 = \{0, 1, 2, 3, 4, 5, 6\}$.   □

Now we have a usable set-theoretic "model" for any particular natural number. But we do not yet have $\mathbb{N}$, because nothing in our axioms warrants the assertion that the collection of all natural numbers forms a set. We will give an axiom that does this indirectly, by asserting the existence of a particular type of infinite set.

A.2.6   DEFINITION

A set X is called an **inductive set** if

1. $\emptyset \in X$.
2. If $A \in X$, then $S(A) \in X$.

AXIOM VII (Axiom of infinity)

There exists an inductive set. □

This does not give us a particular inductive set, in much the same way that the axiom of existence did not by itself assert the existence of any particular set. We will (of course) remedy this by specification. Intuitively, we can see that an inductive set will contain the set 0 and all of its successors. That is, any inductive set must contain every natural number. Furthermore, the set $\mathbb{N}$ should itself be an inductive set. This motivates the following definition.

A.2.7   DEFINITION

The set $\mathbb{N}$ of **natural numbers** is the set of elements contained in every inductive set.

A.2.8   EXERCISE

Show that $\mathbb{N}$ exists and is unique by

1. starting with some inductive set A and constructing $\mathbb{N}$ by specification, and
2. showing that the same $\mathbb{N}$ would result if we had started with another inductive set B. □

A.2.9   THEOREM

$\mathbb{N}$ is an inductive set. □

A.2.10   COROLLARY ("Axiom" of Induction)

Every inductive subset of $\mathbb{N}$ is equal to $\mathbb{N}$. □

$\mathbb{N}$ is thus in some sense the "smallest inductive set." This is exactly the axiom of induction of Chapter 3, expressed in slightly different terms. We can therefore use mathematical induction as a tool to prove other facts about $\mathbb{N}$.

A.2.11   EXERCISE

Explain why we refer to Corollary A.2.10 as the "axiom" of induction. Include in your answer a discussion of two issues:

• Why is the word axiom in quotation marks?
• Why is Corollary A.2.10 the same as the axiom of induction that we discussed in Chapter 3? □

A.2.12   THEOREM

Every natural number $n \in \mathbb{N}$ is either 0 or else the successor of some natural number m: $S(m) = n$. □

A.2.13   DEFINITION

A set X is called **transitive** if, for every $A \in X$, $A \subseteq X$.

This is obviously a different sense of the term "transitive" than the one used in describing a property of a relation on a set. The set-theoretic idea is a strange one: Transitive sets are those whose elements are also subsets. This may make a little more sense after we prove the following theorem.

A.2.14   THEOREM

Every natural number is a transitive set.   □

The fact that natural numbers are transitive is extremely useful in proving the following two technical results, which will in turn help us demonstrate the order structure of $\mathbb{N}$.

A.2.15   THEOREM

For all $n \in \mathbb{N}$, $n \notin n$. (*Hint:* Use induction, together with the fact that n is a transitive set.)   □

A.2.16   THEOREM

For all $k, n \in \mathbb{N}$, if $k \in n$ then, $n \not\subseteq k$. (*Hint:* Phrase this as a predicate about n, and then prove it using induction, making use of the previous theorem.)   □

Now we are in a position to prove that $\mathbb{N}$ is totally ordered. It will make things easier if we first use induction to prove the following lemma.

A.2.17   LEMMA

For each $k \in \mathbb{N}$, define $C_k = k \cup \{n \in \mathbb{N} : k \subseteq n\}$. Show that $C_k = \mathbb{N}$ for all $k \in \mathbb{N}$.   □

A.2.18   THEOREM

$\mathbb{N}$ is totally ordered by the relation $\subseteq$.   □

In our construction of $\mathbb{N}$, the total ordering is just the subset relation $\subseteq$. In the particular context of numbers, we usually use the more familiar symbol $\leq$ for the ordering. Thus, "$m \leq n$" as a statement about natural numbers means "$m \subseteq n$" as a statement about sets.

A.2.19   THEOREM

$\mathbb{N}$ is well-ordered—that is, every nonempty subset of $\mathbb{N}$ has a least element.   □

A.2.20   EXERCISE

Show that every natural number is the set of all previous natural numbers—that is, for all $n \in \mathbb{N}$, $n = \{m \in \mathbb{N} : m < n\}$. (As usual, the expression "$m < n$" means that $m \leq n$ and $m \neq n$.)   □

The structure of $\mathbb{N}$ is sometimes expressed by the *Peano axioms*, a set of elementary statements about natural numbers that form a sufficient basis for all ordinary number theory. Now that we have mastered the set-theoretic construction of $\mathbb{N}$, it should be straightforward to prove each of the Peano axioms.

A.2.21   THEOREM (Peano axioms)

The following facts hold for $\mathbb{N}$.

1. $0 \in \mathbb{N}$.
2. For every $n \in \mathbb{N}$, there exists a successor $S(n) \in \mathbb{N}$.
3. For every $n \in \mathbb{N}$, $0 \neq S(n)$.
4. For every $m, n \in \mathbb{N}$, if $m \neq n$ then $S(m) \neq S(n)$.
5. If $A \subseteq \mathbb{N}$ such that $0 \in A$ and for every $x \in A$, $S(x) \in A$, then $A = \mathbb{N}$.   □

From this point, we can build up the entire structure of natural number arithmetic. In the remainder of this section I give only a brief outline of how this development proceeds. First, I must define the operations of addition and multiplication on $\mathbb{N}$ on the basis of our axioms (either using the Peano axioms for $\mathbb{N}$ or appealing directly to the axioms of set theory). The definitions of the operations turn out to be *recursive* definitions.

A.2.22   DEFINITION

For any $n, m \in \mathbb{N}$, define the natural number $n + m$ as follows:

• $n + 0 = n$.
• For all $k \in \mathbb{N}$, $n + S(k) = S(n + k)$.

This is called the binary operation of **addition** on $\mathbb{N}$.

This definition has many obvious advantages; for instance, it implies the intuitive formula $S(n) = n + 1$ for all $n \in \mathbb{N}$. But it is not yet completely clear on the basis of the axioms that this really *is* a definition, that it is sufficient to specify a function from $\mathbb{N} \times \mathbb{N}$ to $\mathbb{N}$ (which is what a binary operation is). How do we know that the function exists? Justifying our definition requires the use of a powerful result of set theory.

A.2.23   THEOREM (The recursion theorem)

Let A be a set and let f be a function from A into A. Let $a \in A$. Then there exists a function F from $\mathbb{N}$ into A such that

- $F(0) = a$.
- For all $n \in \mathbb{N}$, $F(S(n)) = f(F(n))$.

That is, F is the sequence created by an iterated map of f based at a.   □

Proof Sketch.   This theorem is a bit tricky to prove. The basic strategy goes like this. The function in question, F, is really a subset of the Cartesian product $\mathbb{N} \times A$. In fact, it is a subset with the following properties:

1. $(0, a) \in F$.
2. If $(n, x) \in F$, then $(S(n), f(x)) \in F$.

Of course, F is not the *only* subset of $\mathbb{N} \times A$ with these properties—$\mathbb{N} \times A$ itself has them. But F is the smallest such subset, and it is contained in any subset of $\mathbb{N} \times A$ with these properties. Thus, to prove the recursion theorem, we should construct F by specification as an appropriate subset of $\mathbb{N} \times A$ and then prove (using induction) that F is indeed a function. The details are left to you!   □

Once we accept the recursion theorem, we can use it to justify our definition of addition.

A.2.24   EXERCISE

For each n, use the recursion theorem to define the function f_n from $\mathbb{N}$ into $\mathbb{N}$ by $f_n(m) = n + m$, as specified above. Then use the f_n's to construct the addition function $+$ from $\mathbb{N} \times \mathbb{N}$ into $\mathbb{N}$.   □

With arithmetic operations defined, we can set about proving their properties. Each proof is typically built around a fairly involved induction argument.

A.2.25   PROBLEM

Show that addition, as defined above, is commutative and associative.   □

A.2.26   PROBLEM

Define multiplication on $\mathbb{N}$ recursively. Justify the definition using the recursion theorem, and show that multiplication is commutative and associative.   □

A.2.27   PROBLEM

Show that multiplication distributes over addition in $\mathbb{N}$; that is, for all $k, m, n \in \mathbb{N}$,

$$k(m + n) = km + kn$$

where we have (as is customary) omitted the operation symbol for multiplication.   □

There are also relationships between the arithmetic operations and the total ordering on $\mathbb{N}$. Here is a relatively easy example.

A.2.28   EXERCISE

Let $a, m, n \in \mathbb{N}$. Show that if $m \leq n$, then $m + a \leq n + a$.   □

A.2.29   EXERCISE

Formulate a theorem relating multiplication and total ordering in $\mathbb{N}$, and then prove it.

□

A.3   Axioms of Choice and Substitution

With the axiom of infinity as a starting point, we can construct larger and larger infinite sets, as described in Chapter 7. Such bigger and bigger infinities pose some new puzzles, which set theory must deal with. Consider the following theorem about cardinality.

A.3.1   THEOREM

Let A and B be sets, and suppose $f : A \rightarrow B$ is onto. Then card $B \leq$ card A.   □

This is a very intuitive statement: If A can "cover" B via a function f, then B can be "no bigger" than A. It certainly *ought* to be true, and a countable version of the problem appeared in Section 7.3, where it was useful in proving that the set of rational numbers is countable. What is required to prove this highly desirable theorem?

To show that card $B \leq$ card A, we need to show the existence of a one-to-one function $g : B \rightarrow A$. The function g will be a sort of "inverse" to f. For each element $x \in B$, there is a set $f^{-1}(\{x\}) \subseteq A$, the inverse image of the set $\{x\}$. We will construct g so that $g(x)$ is some element in $f^{-1}(\{x\})$. Which one? It does not matter. For each $x \in B$, choose *any* element of $f^{-1}(\{x\})$ to be $g(x)$. Since the inverse images $f^{-1}(\{x\})$ and $f^{-1}(\{y\})$ are disjoint for distinct x and y, the resulting g will be one-to-one.

This procedure of choosing representatives is clearly a very natural one. If we have a collection of nonempty sets (the various $f^{-1}(\{x\})$'s), it seems reasonable to pick one element from each. The difficulty is that this procedure cannot always be justified by the set-theoretic axioms we have adopted so far. After all, if A and B are infinite, then it may not be possible to give a "step-by-step" algorithm for constructing g.

In order to prove the theorem in general, we will need to adopt a new axiom of set theory that justifies our construction of g. This is the axiom of choice.

A.3.2   DEFINITION

Let X be a nonempty set whose elements are nonempty sets. A **choice function** is a function $h : X \rightarrow \cup X$ such that $h(A) \in A$ for all $A \in X$.

A.3.3   EXERCISE

Explain why a choice function expresses the idea of "choosing a representative" from each member of a collection of nonempty sets.   □

AXIOM VIII (Axiom of choice)

For any nonempty set whose elements are nonempty sets, a choice function exists.   □

A.3.4   EXERCISE

Use the axiom of choice to show the existence of the function g in the suggested proof of Theorem A.3.1.   □

As an immediate consequence of our now-justified proof, we have the following.

A.3.5   EXERCISE

If A and B are sets and $f : A \to B$, then card $f(A) \le$ card A.   □

A.3.6   PROBLEM

The axiom of choice is not required as a new axiom for finite sets. Let X be a finite nonempty set whose elements are nonempty sets. Show *without using the axiom of choice* that there is a choice function on X. (*Hint:* Proceed by induction on the size of X.)   □

You have probably made use of the axiom of choice without realizing it, particularly in Chapter 7. For example, Theorem 7.2.5 asserts that every infinite set contains a sequence of distinct terms. To build such a sequence, one must make an infinite number of "choices." This process can be justified only by the axiom of choice. To see this, try the following slightly easier version of the result.

A.3.7   PROBLEM

Using the set-theoretic axioms explicitly, show that any infinite set contains a countably infinite subset. (*Hint:* Let X be infinite. Show that by definition, X contains subsets of any finite size. Let S_n be the set of subsets of X having n elements. Use the axiom of choice to choose one representative from each S_n, and form from them a countably infinite subset of X.)   □

The axiom of choice was once somewhat controversial. The axiom does not actually specify a particular choice function; it merely asserts that one exists. In other words, the axiom is rather *nonconstructive*. Furthermore, when applied to extremely large sets, the axiom of choice can sometimes produce surprising and nonintuitive results. However, it has proved to be extremely useful in many branches of mathematics and is now a standard part of ordinary set theory.

Some appreciation of the power of the axiom of choice can be obtained by surveying this brief list of mathematical facts that are derived from it. The proofs of some of these statements are quite dfficult and are beyond the scope of this book.

- For any two sets A and B, either card $A \leq$ card B or card $B \leq$ card A. (In other words, the cardinalities of any two sets are comparable.)
- For every infinite set A, $A \times A$ has the same cardinality as A.
- For every infinite set A, the set of finite subsets of A has the same cardinality as A.
- (Zorn's lemma) Let X be a partially ordered set, and suppose that every chain in X has an upper bound. Then X has a maximal element. (A "chain" is a totally ordered subset of the partially ordered set.)
- (Well-ordering principle) Every set can be well ordered. (The ordering here may have nothing to do with some conventional order on the set. $\mathbb{R}$ is certainly not well ordered under the usual ordering!)
- Every vector space has a basis.

Several of these statements, including Zorn's lemma and the well-ordering principle, are in fact logically equivalent to the axiom of choice.

There remains one additional axiom in standard set theory, called the axiom of substitution. We can motivate the new axiom by an informal example. Suppose we have somehow established the existence of the set of Boy Scouts, and we have also established that every Boy Scout has exactly one mother. Then it seems reasonable to posit that the mothers of Boy Scouts also form a set—intuitively, we can just take the set of Boy Scouts and "substitute" for each Boy Scout the corresponding mother.

Here is the necessary axiom.

AXIOM IX (Axiom of substitution)

Suppose $Q(x, y)$ is a predicate (with two free variables) such that for every x there is a unique y for which $Q(x, y)$ is true. Let A be a set. Then there is a set B such that, for every $x \in A$, there exists $y \in B$ for which $Q(x, y)$ is true.   □

A.3.8   EXERCISE

Identify the predicate $Q(x, y)$ and the sets A and B in our informal motivating example.

□

Like the axiom of specification, this is actually an "axiom schema," a rule for generating infinitely many statements about sets (one for each $Q(x, y)$). The axiom of substitution allows us to use predicates of a special sort to build new sets from old. The predicate $Q(x, y)$ works as a sort of "function" mapping elements $x \in A$ to elements y in the new set B for which $Q(x, y)$ holds.

The principal use of the axiom of substitution is to handle extremely large infinite sets. To illustrate this, we will construct a set that is far larger than any other set mentioned in this book.

A.3.9 EXERCISE

Let $T_0 = \mathbb{N}$, $T_1 = \mathcal{P}(\mathbb{N})$, $T_2 = \mathcal{P}(\mathcal{P}(\mathbb{N}))$, and so on. That is, let $T_{n+1} = \mathcal{P}(T_n)$ for each natural number n.

1. Show that the set $T = \{T_0, T_1, \ldots\}$ exists. (*Hint:* Remember, a mere description of a bunch of items does not guarantee that they form a set. Consider the predicate $Q(x, y) := \text{``}y = T_x\text{''}$ and apply the axiom of substitution.)

2. Now consider the set $\cup T$. Prove that $\text{card}(\cup T) > \text{card } T_n$ for all $n \in \mathbb{N}$. In other words, $\cup T$ is larger than $\mathbb{N}$, $\mathcal{P}(\mathbb{N})$, $\mathcal{P}(\mathcal{P}(\mathbb{N}))$, and so on.   □

B Constructing ℝ

In Chapter 8 a set of axioms was assembled to describe the structure of the real number system ℝ. These axioms were the basis for discussing the arithmetic and order properties of ℝ, the existence of irrational numbers such as $\sqrt{2}$, and sequence convergence. But our listing of the real number "axioms" was not really a *construction* of the real numbers. Put another way, listing a set of basic properties for ℝ does not guarantee that any mathematical object satisfying those properties actually exists.

In this appendix, we will demonstrate that in fact such an object does exist. We will construct, on the basis of axiomatic set theory, a mathematical structure satisfying all of the "axioms" for ℝ. This is analogous to our construction, in Appendix A, of a set satisfying all of the Peano axioms for the natural numbers ℕ. The construction of ℝ reinforces the idea that the axioms of set theory are an adequate foundation for all ordinary mathematics.

So this appendix lays the set-theoretic foundation for Chapter 8. The *axioms* for ℝ in that chapter will become *theorems* about the set structure we build here.

Our starting point will be the natural numbers ℕ, whose construction on the basis of the axioms of set theory was outlined in Appendix A.[1] We will assume the principle of induction, the total ordering of ℕ, and the arithmetic operations of addition and multiplication. All of the structure of ℝ will be built out of these basic ingredients.

Instead of constructing ℝ at once, our strategy will be to build it up in stages, by constructing two intermediate structures: the set of integers ℤ and the set of rational numbers ℚ. That is, we will proceed as follows:

$$\mathbb{N} \longrightarrow \mathbb{Z} \longrightarrow \mathbb{Q} \longrightarrow \mathbb{R}.$$

---

[1] Of course, we mean the *set-theoretic* natural numbers ℕ, which includes the number 0.

Since we already have a lot of informal knowledge about $\mathbb{Z}$, $\mathbb{Q}$, and $\mathbb{R}$, we will be able to use this intuition as a guide for our constructions at each stage. We are after all trying to use the materials of set theory to build a mathematical structure with particular characteristics, so it is okay to "look ahead" and pick our definitions to suit our eventual goal. But even though we may anticipate our final result when making definitions, we will be careful in our proofs to use only previously established results.

B.1   From ℕ to ℤ

In this section we will use the natural numbers $\mathbb{N}$ to construct the integers $\mathbb{Z}$, which include both positive and negative numbers. Why are we interested in negative numbers? The answer is, because we wish to have available the operation of *subtraction*. (In a similar way, we will construct the rational numbers $\mathbb{Q}$ in the next section because we wish to be able to perform the operation of *division*.)

We will find the following results about $\mathbb{N}$ to be useful.

B.1.1   THEOREM

1. If $a, b, c \in \mathbb{N}$, then $a + c = b + c$ if and only if $a = b$. (*Hint:* Prove this by induction on c, using the recursive definition of addition on $\mathbb{N}$ from Appendix A.)

2. If $a, b \in \mathbb{N}$, and $a \geq b$, then there exists a unique $c \in \mathbb{N}$ such that $a = b + c$. (*Hint:* This is a "double induction" proof. First fix $b = 0$ and prove the statement using induction on a; then do induction on b.)

3. If $a, b, c \in \mathbb{N}$, and $c \neq 0$, then $a \cdot c = b \cdot c$ if and only if $a = b$. □

The first and third facts are "cancellation" laws for addition and multiplication in $\mathbb{N}$. In an equation of natural numbers in which c has been added to each side (or, if $c \neq 0$, multiplied by each side), it is correct to "cancel" the c term on both sides. This is like "subtracting c from both sides" (or "dividing both sides by c"). The second fact comes even closer to the idea of subtraction. We could interpret the expression "$12 - 5$" to be "the unique natural number c such that $12 = 5 + c$." Unfortunately, this only works because $12 \geq 5$; we cannot make sense within $\mathbb{N}$ of the expression "$5 - 12$."

Our idea, therefore, is that an integer is given by the subtraction of one natural number from another. Integers will be constructed out of *ordered pairs* of natural numbers. But different subtractions can lead to the same integer. If an integer is indeed a subtraction, we would want all subtractions that lead to the same integer to be considered equivalent. When does $a - b = c - d$? Add $b + d$ to both sides!

B.1.2   THEOREM

Consider the relation $\approx$ on the Cartesian product $\mathbb{N} \times \mathbb{N}$, defined by

$$(a, b) \approx (c, d) \text{ if and only if } a + d = b + c.$$

Then $\approx$ is an equivalence relation.

(*Hint:* You may not use subtraction to prove this theorem! Since we have not yet developed the machinery that will allow us to subtract one natural number from another, you must rely solely on the additive properties of ℕ.)   □

B.1.3 DEFINITION

An equivalence class of pairs of natural numbers under the relation ≈ is called an **integer.** Denote the equivalence class of (a, b) by $[a - b]$, and the set of equivalence classes by ℤ.

We've chosen the notation $[a - b]$ for the equivalence class containing (a, b) to emphasize our intuitive motivation: An integer is a "difference" of natural numbers. Nevertheless, for now the symbol "−" is merely part of our notation for the equivalence classes under ≈ and does not represent an operation.

It is possible to construct ℤ in other ways. For example, we could have simply added to ℕ a set of "negative integers," defining their properties as needed. Our construction is designed to be as parallel as possible to the construction of ℚ in the next section.

B.1.4 EXERCISE

Show $(a, b) \approx (a + c, b + c)$ for all $a, b, c \in ℕ$.   □

B.1.5 EXERCISE

Describe the integers $[1 - 1]$, $[4 - 1]$, and $[0 - 3]$.   □

We can also demonstrate this important result:

B.1.6 THEOREM

Suppose $a, b \in ℕ$.

- If $a > b$, then $(a, b) \in [n - 0]$ for some unique $n \in ℕ$ with $n > 0$.
- If $a = b$, then $(a, b) \in [0 - 0]$.
- If $a < b$, then $(a, b) \in [0 - n]$ for some unique $n \in ℕ$ with $n > 0$.   □

This means that the pairs of the form $(n, 0)$, $(0, 0)$, and $(0, n)$ where n ranges over all nonzero natural numbers form a *complete set of equivalence class representatives* in the sense of Chapter 6. Each equivalence class has exactly one representative of this form.

B.1.7 DEFINITION

We introduce the following relations and operations on ℤ.

- An **order** relation ≤ on ℤ is defined by: $[a - b] \le [c - d]$ if and only if $a + d \le b + c$.
- A **negation** function from ℤ to ℤ is defined by $-[a - b] = [b - a]$.
- A binary operation of **addition** on ℤ is defined by $[a - b] + [c - d] = [(a + c) - (b + d)]$.
- A binary operation of **multiplication** on ℤ is defined by $[a - b] \cdot [c - d] = [(ac + bd) - (ad + bc)]$.

Note that we specified the same symbols for the relations and operations on $\mathbb{Z}$ that we use for corresponding relations and operations on $\mathbb{N}$. For any given symbol, context will tell us which meaning is intended.

B.1.8 QUESTION

If we think of the integer $[a - b]$ as the result of the subtraction $a - b$, why is each definition above a reasonable one?

Note that we have defined the various relations and operations on $\mathbb{Z}$ by means of *representatives* of the equivalence classes. This means that, for each definition above, we must resolve the key issue of *well-definedness*. Each definition refers to one or more representatives, but the resulting relation or operation had better not depend on *which particular* representatives are used. A similar issue arose in Chapter 6 when discussing arithmetic modulo n, and in the discussion of comparing cardinalities in Chapter 7.

B.1.9 THEOREM

Show that the relation $\le$ on $\mathbb{Z}$ is well-defined. (*Hint:* Start by supposing $(a, b) \approx (a', b')$, so that (a, b) and (a', b') are in the same equivalence class. Do the same for pairs (c, d) and (c', d'). Now show that $a + d \le b + c$ if and only if $a' + d' \le b' + c'$.) □

B.1.10 THEOREM

Show that negation, addition, and multiplication on $\mathbb{Z}$ are each well-defined. □

Now that we have well-defined relations and operations, we can prove their properties.

B.1.11 THEOREM

Show that $\le$ is a total ordering on $\mathbb{Z}$. □

B.1.12 EXERCISE

Suppose that $x, y \in \mathbb{Z}$. Show that, if $x \le y$, then $-y \le -x$. □

B.1.13 THEOREM

Show that addition on $\mathbb{Z}$ is commutative and associative. □

B.1.14 THEOREM

Show that, for every $x \in \mathbb{Z}$, $x + (-x) = [0 - 0]$. □

B.1.15 THEOREM

Show that multiplication on $\mathbb{Z}$ is commutative and associative. □

Without warning we have begun to refer to elements of $\mathbb{Z}$ using conventional notation. We will do this whenever possible—that is, whenever it is not necessary to refer to a *specific pair* in the equivalence class that gives the integer. As you would hope, by the end of this section, we will have ceased entirely to refer directly to the equivalence classes, and we will be able to handle integers in an "everyday manner."

B.1.16 THEOREM

Show that multiplication is distributive over addition in $\mathbb{Z}$.   □

We now turn to a very important point. Informally, we usually think of the natural numbers $\mathbb{N}$ as a *subset* of the integers $\mathbb{Z}$. This is not so in our construction of $\mathbb{Z}$. Instead, we will show that $\mathbb{N}$ may be "embedded" in $\mathbb{Z}$; or, in other terms, that $\mathbb{Z}$ contains a subset that is *isomorphic* to $\mathbb{N}$.[2] From our previous discussions of isomorphism, you will recall that two mathematical structures are isomorphic if the difference between them amounts to a relabling of the elements. We further said that this idea can be made mathematically rigorous by constructing a structure-preserving one-to-one correspondence between them. In the case of partially ordered sets, we spoke of order isomorphisms. In this case, we will need a one-to-one function from $\mathbb{N}$ into $\mathbb{Z}$ that preserves all the important mathematical properties of $\mathbb{N}$: the order structure, the additive structure, and the multiplicative structure.

B.1.17 THEOREM

Define a function $e : \mathbb{N} \to \mathbb{Z}$ by $e(n) = [n - 0]$ for all $n \in \mathbb{N}$. Let $a, b \in \mathbb{N}$. Then e has the following properties.

1. e is one-to-one.
2. $e(a) \le e(b)$ if and only if $a \le b$. (*This means that e is an order isomorphism in the sense of Section 5.4, and thus the order structure of* $\mathbb{N}$ *is preserved by e.*)
3. Show that $e(a + b) = e(a) + e(b)$. (*This part shows that the additive structure of* $\mathbb{N}$ *is preserved by e.*)
4. Show that $e(ab) = e(a) \cdot e(b)$. (*This part shows that e preserves the multiplicative structure of* $\mathbb{N}$.)   □

In other words, the image $e(\mathbb{N})$ is a subset of $\mathbb{Z}$ in one-to-one correspondence to $\mathbb{N}$ with exactly the same order and arithmetic structure. For practical purposes, we can pretend that $e(\mathbb{N})$ "is" $\mathbb{N}$. We can just write "n" instead of "$[n - 0]$" for the image in $\mathbb{Z}$ of some natural number n. This is a sensible but slightly confused notation. Bear in mind that the natural number n is not the same as the integer n (which is really an equivalence class of ordered pairs of natural numbers) though we will talk as if they are the same object.

B.1.18 EXERCISE

Show that every integer can either be written as n or $-n$ for a natural number n, and that $-0 = 0$.   □

---

[2] You may recall that we did something similar in Chapter 8 when we found a "copy" of $\mathbb{N}$ inside (the axiomatically assumed) $\mathbb{R}$.

B.1.19   EXERCISE

Prove the following facts about arithmetic in $\mathbb{Z}$. Let $a, b, c \in \mathbb{Z}$.

1. $a + 0 = a$.
2. $a \cdot 0 = 0$.
3. $a \cdot 1 = a$.
4. $-1 \cdot a = -a$.
5. If $a + c = b + c$, then $a = b$.
6. If $c \neq 0$ and $a \cdot c = b \cdot c$, then $a = b$.   □

We may exploit our confused notation (in which "$\mathbb{N} \subseteq \mathbb{Z}$") to illuminate the definition of the integers.

B.1.20   EXERCISE

Define the binary operation of **subtraction** by $a - b = a + (-b)$ for all $a, b \in \mathbb{Z}$. Show that if $m, n \in \mathbb{N}$ then $m - n = [m - n]$.   □

For the rest of this appendix, we will assume familiar facts about integer addition and multiplication as necessary. All of them follow from our construction of $\mathbb{Z}$ and can be proven if the need arises.

B.2   From $\mathbb{Z}$ to $\mathbb{Q}$

The development of the rational numbers $\mathbb{Q}$ from $\mathbb{Z}$ will closely parallel the development of $\mathbb{Z}$ from $\mathbb{N}$ in the previous section. Our construction of the integers was motivated by the idea that an integer was the *difference* of two natural numbers; thus we defined integers as equivalence classes in $\mathbb{N} \times \mathbb{N}$. Now we will think of each rational number as a *quotient* of integers.

B.2.1   THEOREM

Let Ψ be the subset of $\mathbb{Z} \times \mathbb{Z}$ such that $(a, b) \in \Psi$ if $b \neq 0$. Define a relation $\sim$ on Ψ by $(a, b) \sim (c, d)$ if and only if $ad = bc$. Show that $\sim$ is an equivalence relation.[3]

(*Hint:* You may not use division of integers to prove this theorem! Remember, we have not yet developed the machinery that will allow us to divide one integer by another.)

□

B.2.2   DEFINITION

The equivalence classes of Ψ are called **rational numbers.** The set of rational numbers is denoted $\mathbb{Q}$, and the equivalence class containing (a, b) is denoted $[a/b]$.

---

[3] This theorem appeared as Problem 17 at the end of Chapter 4.

B.2.3   QUESTION

Why do we exclude pairs of the form $(x, 0)$ from the set Ψ?

Remark.   This question should not be answered with a facile, "We can't divide by zero." It is trying to get at the fundamental issue of *why* we can't divide by zero. This fact is not arbitrary and must therefore be inherent in the mathematics itself. Go back to your proof of Theorem B.2.1 and see where exactly you used the fact that no pairs of the form $(x, 0)$ were present.

B.2.4   EXERCISE

Describe the equivalence classes $[0/-1]$ and $[2/3]$.   □

We can begin to build the structure of $\mathbb{Q}$ in a way similar to that we used in constructing $\mathbb{Z}$.

B.2.5   DEFINITION

We introduce the following functions and operations on $\mathbb{Q}$.

- A **negation** function from $\mathbb{Q}$ to $\mathbb{Q}$, is defined by $-[a/b] = [(-a)/b]$.
- A **reciprocal** function defined on a subset of $\mathbb{Q}$, is defined for $[a/b]$ with $a \neq 0$ by $[a/b]^{-1} = [b/a]$.
- A binary operation of **addition** is defined by $[a/b] + [c/d] = [(ad + bc)/(bd)]$.
- A binary operation of **multiplication** is defined by $[a/b] \cdot [c/d] = [(ac)/(bd)]$.

Once again, we have employed many of the same symbols used in the integers for relations and operations on $\mathbb{Q}$, since the meaning of each symbol will be clear from the context.

B.2.6   QUESTION

Why is the reciprocal function only defined on a *subset* of $\mathbb{Q}$?

Naturally, since the above definitions refer to specific representatives of the equivalence classes, we need to prove that they are well-defined.

B.2.7   PROBLEM

Prove that each of the functions and operations given in Definition B.2.5 is well-defined.   □

We can then begin to verify the arithmetic properties of $\mathbb{Q}$, which are summarized in the *field axioms*—for us, theorems based upon our construction of $\mathbb{Q}$.

B.2.8   THEOREM

Show that addition on $\mathbb{Q}$ is commutative and associative.   □

B.2.9   THEOREM

Show that ℚ contains an additive identity and additive inverses.

- For any $x \in ℚ$, $x + [0/1] = x$.
- For any $x \in ℚ$, $x + (-x) = [0/1]$.   □

B.2.10   THEOREM

Show that multiplication on ℚ is commutative and associative.   □

> Here, as in Section B.1, we move away from our equivalence class notation for the rationals toward more standard notation as soon as enough structure is built to make it possible.

B.2.11   THEOREM

Show that ℚ contains a multiplicative identity and multiplicative inverses.

- For any $x \in ℚ$, $x \cdot [1/1] = x$.
- For any $x \in ℚ$ such that $x \neq [0/1]$, $x \cdot x^{-1} = [1/1]$.   □

B.2.12   THEOREM

Show that addition distributes over multiplication in ℚ.   □

Notice that we have not yet discussed the total ordering of ℚ. In Chapter 8 order structure was introduced via the "order axiom." The order axiom asserted the existence of a special subset of a field, called the "positive" subset, which was closed under addition and multiplication.

B.2.13   DEFINITION

Define a set ℚ⁺ of **positive rational numbers** by

$$ℚ^+ = \{[a/b] \in ℚ : \text{ both } a \text{ and } b \text{ are greater than } 0\}.$$

B.2.14   EXERCISE

Show that ℚ⁺ is well-defined.   □

We can prove the following.

B.2.15   THEOREM (Order "Axiom")

ℚ⁺ is closed under addition and multiplication. That is, if $x, y \in ℚ^+$, then $x + y \in ℚ^+$ and $x \cdot y \in ℚ^+$.   □

B.2.16   PROBLEM

Define a relation $\leq$ on ℚ as follows: For $x, y \in ℚ$, $x \leq y$ if and only if either $x = y$ or $(-x) + y \in ℚ^+$. Show that $\leq$ is a total ordering on ℚ.   □

B.2.17   PROBLEM

Devise an alternative definition of the relation $\leq$ in the style of the definitions of negation, addition, etc.—as a relation based on integer properties of representative pairs. Show that your definition is well-defined, and that it gives the same total ordering for ℚ.   □

We have now shown that ℚ is an "ordered field," whose arithmetic properties are familiar. Consult Chapter 8 for various arithmetic facts that can be proved from the field axioms and the order axiom.

There remains the problem of "embedding" the integers ℤ inside the rational numbers ℚ. After all, we usually think of the integers ℤ as a special subset of ℚ, not just as a more basic structure out of which we construct ℚ. What this means is that ℚ contains a subset that is *isomorphic* to ℤ, in much the same way that ℤ contained a subset that was isomorphic to ℕ.

B.2.18   THEOREM

Define a function $f : \mathbb{Z} \to \mathbb{Q}$ by $f(n) = [n/1]$ for all $n \in \mathbb{Z}$. Let $a, b \in \mathbb{Z}$.

1. Show that f is one-to-one.
2. Show that $f(a) \leq f(b)$ if and only if $a \leq b$.
3. Show that $f(-a) = -f(a)$.
4. Show that $f(a + b) = f(a) + f(b)$.
5. Show that $f(ab) = f(a) \cdot f(b)$.   □

Once again, we adopt the standard practice of considering $f(\mathbb{Z})$ to "be" ℤ itself, and refer to the rational number of the form $f(a) = [a/1]$ for integer a simply as "a." (The natural numbers, too, are embedded in ℚ in the set $f(e(\mathbb{N}))$.) Adopting this convention leads us to the following exercise, which makes rigorous the idea that a rational number is a "quotient of integers."

B.2.19   EXERCISE

Define the operation of **division** on ℚ by $x \div y = x \cdot (y^{-1})$. (Note that this is only defined if $y \neq 0$.) If $m, n \in \mathbb{Z}$, show that $m \div n = [m/n]$.   □

Finally we can adopt the conventional fraction notation $[a/b] = \frac{a}{b}$ for rational numbers.

B.3   From ℚ to ℝ

Now we are ready to use the rational numbers ℚ to construct the real numbers ℝ. As in the previous two stages, we will choose our definitions so that we can perform an operation in ℝ that is not always possible in ℚ. The discussion in Chapter 8 suggests

Figure B.1   A Dedekind cut

that this new operation should be finding the *least upper bound* of a set of rationals that is bounded above.

Thus, we might define an equivalence relation on the subsets of ℚ that are bounded above: $A \asymp B$ if and only if A and B have the same upper bounds in ℚ. Real numbers would then be identified with the equivalence classes given by this relation, and the various relations, functions, and operations on ℝ would be defined using representatives of the equivalence classes. (We would take care, of course, that each new object was well-defined.) The rationals ℚ would be found imbedded within ℝ in some natural way.

However, this construction brings with it certain technical difficulties that can be avoided by pursuing instead a different (but equivalent) construction based upon *particular* subsets of ℚ that are bounded above. This will greatly simplify our task in building the various relations, functions, and operations on ℝ (particularly in proving the order properties). The construction we follow was first given by Richard Dedekind in 1872.

An equivalent construction was proposed by Cantor at about the same time. Cantor's version of ℝ was built out of equivalence classes of *Cauchy sequences* in ℚ. The resulting space would therefore be one in which every Cauchy sequence of rational numbers was guaranteed to converge to something in ℝ. See Section 8.5.

B.3.1   DEFINITION

A **Dedekind cut** x of ℚ is a nonempty proper subset of ℚ such that

1. If $a \in x$ and $b \in ℚ$ such that $b < a$, then $b \in x$.
2. If $a \in x$, then there is a $c \in x$ for which $a < c$.

Intuitively, a Dedekind cut is an "initial segment" of ℚ, as shown in Figure B.1. It is useful to notice some properties of Dedekind cuts.

B.3.2   EXERCISE

Two elementary facts:

1. Show that a Dedekind cut x has no greatest element.
2. Show that a Dedekind cut x has an upper bound.   □

B.3.3   EXERCISE

Let $a \in ℚ$. Show that the set $\{b \in ℚ : b < a\}$ is a Dedekind cut.   □

B.3.4   QUESTION

Are all Dedekind cuts of the form given in Exercise B.3.3? Why or why not?

By now you have probably guessed our next move.

B.3.5   DEFINITION

A Dedekind cut x of the rationals $\mathbb{Q}$ is called a **real number.** The set of real numbers is denoted by $\mathbb{R}$.

We should thus think of real numbers as places where we can "break" $\mathbb{Q}$ into two pieces, so that the left-hand piece is a Dedekind cut. The break-point may either be at a rational number (as in Exercise B.3.3) or "between" rational numbers. In this way the set of real numbers includes both rational points and irrational points.

Let us formally define the use of the terms "rational" and "irrational" to describe Dedekind cuts (real numbers).

B.3.6   DEFINITION

Suppose x is a Dedekind cut of $\mathbb{Q}$.

- We say that x is a **rational cut** if there exists $a \in \mathbb{Q}$ such that $b \in x$ if and only if $b < a$.
- We say that x is an **irrational cut** if it is not a rational cut.

We can map elements in $\mathbb{Q}$ to rational Dedekind cuts according to the obvious function: For $a \in \mathbb{Q}$, a^* is the Dedekind cut such that $b \in a^*$ if and only if $b < a$. (Eventually, this function will allow us to "embed" $\mathbb{Q}$ within $\mathbb{R}$. We will return to this below.) Of all the various properties of $\mathbb{R}$, the easiest to show is the order property.

> Notice that the notation a^* used here is parallel to the use of $e(n)$ and $f(n)$ in the constructions of $\mathbb{Z}$ and $\mathbb{Q}$.

B.3.7   THEOREM

Show that $\mathbb{R}$ is a totally ordered set under the relation $\subseteq$. (*Hint:* Since $\subseteq$ is a partial ordering on any system of sets, you will only have to prove that any two Dedekind cuts are related by $\subseteq$.)   □

We will represent the relation $\subseteq$ on $\mathbb{R}$ with the more conventional symbol $\leq$. We can define positive real numbers as follows.

B.3.8   DEFINITION

The set $\mathbb{R}^+$ of **positive real numbers** is defined to be

$$\mathbb{R}^+ = \{x \in \mathbb{R} : 0^* \leq x \text{ and } 0^* \neq x\}.$$

B.3.9   EXERCISE

Show that the following are equivalent definitions for $\mathbb{R}^+$.

$$\{x \in \mathbb{R} : 0 \in x\} \quad \text{and} \quad \{x \in \mathbb{R} : x \cap \mathbb{Q}^+ \neq \emptyset\} \qquad \square$$

Without further ado, we can prove that $\mathbb{R}$ has the least upper bound property.

B.3.10   THEOREM (Least upper bound property)

Suppose $A \subseteq \mathbb{R}$ such that A is bounded above—that is, there is some $x \in \mathbb{R}$ such that $y \leq x$ for all $y \in A$. Then there is a least upper bound for A in $\mathbb{R}$, denoted lub A. (*Hint:* Consider $\cup A$.) $\qquad \square$

The operations of addition and multiplication (and the negation and reciprocal functions) require a little more care. Addition is the easiest.

B.3.11   DEFINITION

The operation of **addition** is defined on $\mathbb{R}$ as follows. Let $x, y \in \mathbb{R}$. Then $x + y$ is the set of all $c \in \mathbb{Q}$ such that $c = a + b$ for some $a \in x$ and $b \in y$.

B.3.12   EXERCISE

For any $x, y \in \mathbb{R}$, show that $x + y \in \mathbb{R}$—that is, that $x + y$ as defined above is itself a Dedekind cut. $\qquad \square$

B.3.13   THEOREM

Addition on $\mathbb{R}$ is commutative and associative. Furthermore, 0^* is an additive identity on $\mathbb{R}$: For all $x \in \mathbb{R}$, $x + 0^* = x$. $\qquad \square$

B.3.14   THEOREM

The set $\mathbb{R}^+$ is closed under addition. That is, if $x, y \in \mathbb{R}^+$, then $(x + y) \in \mathbb{R}^+$. $\qquad \square$

Defining negation is a more subtle business. First, we construct a useful set. Suppose $x \in \mathbb{R}$. Let

$$\overline{x} = \{a \in \mathbb{Q} : (a - r) \notin x \text{ for some } r \in \mathbb{Q}, r > 0\}.$$

In other words, $\overline{x}$ is the set of all rationals that are greater than everything in x by some rational margin r. $\overline{x}$ is not a Dedekind cut, of course; it is a *final* segment rather than an initial one. This suggests the following definition.

B.3.15   DEFINITION

The **negation** of x, denoted $-x$, is just

$$-x = \{a \in \mathbb{Q} : -a \in \overline{x}\}.$$

B.3.16 EXERCISE

Show that $-x \in \mathbb{R}$ for all $x \in \mathbb{R}$. Also show that $-(-x) = x$ for all $x \in \mathbb{R}$.   □

B.3.17 THEOREM

For each $x \in \mathbb{R}$, $x + (-x) = 0^*$. (Thus, every element of $\mathbb{R}$ has an additive inverse.)   □

B.3.18 THEOREM

For every $x \in \mathbb{R}$, exactly one of the following is true.

- $x \in \mathbb{R}^+$
- $x = 0^*$
- $-x \in \mathbb{R}^+$ □

We now turn to the definition of multiplication. It is easiest to define multiplication first for $\mathbb{R}^+$, and then to extend this definition to all of $\mathbb{R}$.

B.3.19 DEFINITION

If $x, y \in \mathbb{R}^+$, then we define $x \cdot y$ to be the set of all $c \in \mathbb{Q}$ such that $c < ab$ for some $a \in x$ and $b \in y$ for which $a > 0$ and $b > 0$.

B.3.20 EXERCISE

Show that $x \cdot y \in \mathbb{R}^+$ for $x, y \in \mathbb{R}^+$. (*Hint:* Don't forget to show that $x \cdot y$ is a Dedekind cut!)   □

B.3.21 DEFINITION

Define the operation of **multiplication** on $\mathbb{R}$ as follows. Let $x, y \in \mathbb{R}$.

- If $x, y \in \mathbb{R}^+$, then $x \cdot y$ is defined as above.
- If $x = 0^*$ or $y = 0^*$, then $x \cdot y = 0^*$.
- If $x \in \mathbb{R}^+$ and $-y \in \mathbb{R}^+$, then $x \cdot y = -(x \cdot (-y))$.
- If $-x \in \mathbb{R}^+$ and $y \in \mathbb{R}^+$, then $x \cdot y = -((-x) \cdot y)$.
- If $-x, -y \in \mathbb{R}^+$, then $x \cdot y = (-x) \cdot (-y)$.

B.3.22 THEOREM

Multiplication on $\mathbb{R}$ is commutative and associative. Furthermore, 1^* is a multiplicative identity on $\mathbb{R}$: For all $x \in \mathbb{R}$, $x \cdot 1^* = x$.   □

B.3.23 THEOREM

Addition distributes over multiplication in $\mathbb{R}$.   □

B.3.24 PROBLEM

Devise a definition for the reciprocal x^{-1} of $x \in \mathbb{R}$ such that $x \neq 0^*$. Show that x^{-1} is the multiplicative inverse for x. ☐

Once we have all of the arithmetic structure of ℝ in place, we can verify that the rational cuts in ℝ are essentially identical to the rational numbers ℚ.

B.3.25 PROBLEM

Show that the * function embeds ℚ in ℝ—that is, that ℚ* (the image of ℚ under the function *) is an isomorphic copy of ℚ. ☐

Following our earlier practice, we drop the * when referring to a rational cut, and simply say $0 \in \mathbb{R}$ (for example) instead of $0^* \in \mathbb{R}$.

We have now verified that the set ℝ of Dedekind cuts of ℚ satisfies all of the axioms given in Chapter 8 for the real number system. Thus, we have established that the standard axioms for set theory are powerful enough to define the real numbers.

It is worth reviewing the major phases of the construction of ℝ. In Appendix A, natural numbers were defined to be transitive sets built up from ∅ by means of the successor operation. In this appendix, integers were defined as equivalence classes of ordered pairs of natural numbers, and rational numbers were defined as equivalence classes of ordered pairs of integers. Finally, the real numbers were identified as special subsets (Dedekind cuts) of the rational numbers. At each stage, arithmetic operations were defined in terms of earlier operations or else (in the case of ℕ) constructed using the recursion theorem. Also earlier structures were found to be isomorphically embedded within later ones.

Our complex and multi-layered construction means that the structure of ℝ is a very "deep" one. The set-theoretic underpinnings of the real number system are quite well concealed! Nevertheless, in some abstract sense the underpinnings are always "there." This means that every proof of a theorem about ℝ—for example, the proof that every Cauchy sequence in ℝ converges—is at some level a set-theoretic argument of dizzying complexity. That valid chains of reasoning of such tremendous size are possible at all is a striking testimony to the power of axiomatic mathematics.

Of course, there is no reason to stop once we have constructed ℝ!

B.3.26 PROBLEM

Construct the complex numbers. ☐

Index

Note: Numbers in **bold** refer to pages on which the terms are defined.